Synthesis Lectures on Digital Circuits & Systems

Series Editor

Mitchell A. Thornton, Southern Methodist University, Dallas, USA

This series includes titles of interest to students, professionals, and researchers in the area of design and analysis of digital circuits and systems. Each Lecture is self-contained and focuses on the background information required to understand the subject matter and practical case studies that illustrate applications. The format of a Lecture is structured such that each will be devoted to a specific topic in digital circuits and systems rather than a larger overview of several topics such as that found in a comprehensive handbook. The Lectures cover both well-established areas as well as newly developed or emerging material in digital circuits and systems design and analysis.

Tyler Kerr

3D Printing

Introduction to Accessible, Affordable
Desktop 3D Printing

 Springer

Tyler Kerr
Innovation Wyrkshop
University of Wyoming
Laramie, WY, USA

ISSN 1932-3166   ISSN 1932-3174 (electronic)
Synthesis Lectures on Digital Circuits & Systems
ISBN 978-3-031-19352-1   ISBN 978-3-031-19350-7 (eBook)
https://doi.org/10.1007/978-3-031-19350-7

This Springer imprint is published by the registered company Springer Nature Switzerland AG
The registered company address is: Gewerbestrasse 11, 6330 Cham, Switzerland

Preface

This book, "3D Printing: Introduction to Accessible, Affordable Desktop 3D Printing," focuses on the foundational concepts of desktop 3D printing and guides readers on a journey to learn to use and even master desktop 3D printers at home. The book starts with an overview of 3D printing, including the general history of the subject, the most common categories of 3D printing, and a number of fascinating current applications across a wide range of industries. Next, we discuss some of the strengths and weaknesses of the most common type of desktop 3D printing—fused deposition modeling (FDM) 3D printing—and review how FDM 3D printers typically work. We explore several popular FDM 3D printer brands and highlight the core features and applications of some of the most common 3D printing materials. We then dive straight into using and mastering 3D printing software, preparing to 3D print, and—perhaps most exciting—finally hitting *Print*. The book culminates with a variety of "STEAM-building" exercises that aim to ignite that innovative spark and get readers thinking about ways in which they might apply 3D printing to their own interests, hobbies, and work.

Approach of the Book

This book is the first in what I hope is a long line of instructional 3D printing books to teach novice readers the ins and outs of many different popular types of 3D printing technology. It should be accessible and applicable to readers with any size budget and any level of preexisting knowledge of 3D printing and emergent technology. Ultimately, the book aims to be approachable enough that anyone—from students, educators, hobbyists, and entrepreneurs—might learn valuable information in each chapter. After all, 3D printers are not just for tech-savvy users or engineers. 3D printing is an innovative and rapidly evolving technology that anyone can master.

This book is divided into ten chapters that aim to give you all the tools and confidence necessary to start 3D printing on your own at home. Though we focus specifically on the popular Prusa i3 MK3S+ desktop FDM 3D printer in many of our examples and lessons, the concepts we'll cover together can be applied to countless other brands of desktop FDM 3D printers.

Chapter 1 covers the basics of 3D printing itself, including broadly how it works and how it has evolved since the 1970s and 1980s. We then explore several different categories of 3D printing technology, including the popular FDM 3D printing, as well as other types of 3D printing you might bump into in your journey, such as stereolithography (SLA), selective laser sintering (SLS), and material jetting (MJ).

There's a common misconception that 3D printers are only for hobbyists, but this is far from the truth! Today, 3D printing can be used in vast and varied ways across almost every industry. Chapter 2 reviews innovative and exciting ways that 3D printing is used today in manufacturing, education, research, art, fashion, food and nutrition, healthcare, housing, automotive work, and the aerospace industry.

For the rest of the book, we hunker down with FDM 3D printers, as these are the most common and thus the most likely 3D printers you'll have access to as you start 3D printing.

Before diving into the specifics of how FDM 3D printing works, we run through a top-down overview of some of FDM 3D printing's strengths and weaknesses. Chapter 3 reviews some of the key features that make FDM 3D printing so popular, and wraps up with a discussion of some of the challenges and limitations of the technology.

Chapter 4 explores FDM 3D printing in greater depth and reviews the typical workflow for getting started, including very basic FDM 3D printer anatomy.

Chapter 5 then reviews a number of popular FDM 3D printer brands, focusing on their strengths and limitations compared to others on the list. Some readers may find Prusa i3 MK3S+ 3D printers, which we discuss throughout the book, an ideal choice. Other readers may have projects that call for an Ultimaker, Creality, LulzBot, Flashforge, or Monoprice 3D printer. Finally, we wrap up the chapter with detailed information on the parts and components specific to the Prusa i3 MK3S+ 3D printer. The concepts learned in Chap. 5, however, should apply to most 3D printer makes and models.

Then, Chap. 6 covers typical filament thermoplastic that you're likely to use as you launch into 3D printing. We'll discuss the benefits and considerations of each type of material, as well as what each material may be well suited for.

Chapters 7 and 8 take a detailed look at how to operate slicing software. Chapter 7 covers the core concepts and four primary features to consider when starting to 3D print. Chapter 8 expands on that knowledge and explores a majority of the more advanced settings you might consider adjusting as you refine and improve your 3D printed project. We then wrap up Chap. 8 with a review of many common problems you're likely to run into, as well as how to start troubleshooting them.

In Chap. 9, it's finally time to print! First, we run through the final steps necessary to prepare a sliced project to print on a Prusa i3 MK3S+. We discuss best practices and final checklists to review before you officially hit *Print*. Then, we hit *Print* and start our project!

Finally, we end the book with Chap. 10, where we cover a wide variety of fun and engaging "STEAM-building" exercises. These projects are intentionally varied and open-ended in nature, and designed to get you thinking about how you might apply what you've learned about 3D printing to your own interests, hobbies, and work.

Laramie, WY, USA   Tyler Kerr
August 2022

Acknowledgements

I want to thank Dr. Steve Barrett, who fostered and encouraged me early in my makerspace career and, more recently, in my first foray into book writing. Despite the best efforts from my mother and father—both editors and authors—writing books was never on my radar. Without Steve, this wonderful opportunity would not have been possible. I also want to thank Joel Claypool of Morgan and Claypool Publishers for taking a chance on this topic and on a new author. I hope ours will be a long and productive partnership. I would also like to thank Dharaneeswaran Sundaramurthy and Vidyalakshmi Velmurugan from Springer Nature for their help and patient guidance in turning this draft into a polished, finished book. I must also thank Prusa Research and Ultimaker for making the 3D printing process approachable and accessible, and for their permission to use hardware and software imagery throughout this book. I would also like to thank AI SpaceFactory, Alquist 3D, the D'Arcy Thompson Zoology Museum at the University of Dundee, Dani Clode, e-NABLE, the Mace Brown Museum of Natural History at the College of Charleston, the Naturalis Biodiversity Center, PolyMaker, and Tel Aviv University for their consent to use images of their products, projects, research, and copyrighted material in this book.

Finally, I would be remiss not to mention my family, Mary, John, Cam, and Natasha, and my ever-supportive partner Kate—all of whom patiently endured many long, overly excited stream-of-consciousness rants about 3D printing over the years. Last, I should thank Addi and Murphy, our loveable but attention-starved dogs. Without you two, this book might have been submitted ahead of schedule, and with far fewer squeaky toy interruptions.

Contents

1 3D Printing 101 1
  1.1 Objectives 1
  1.2 Overview 1
  1.3 What is 3D Printing? 2
  1.4 History of 3D Printing 3
  1.5 Common Categories of 3D Printing 4
     1.5.1 Fused Deposition Modeling (FDM/FFF) 5
     1.5.2 Stereolithography (SLA) 6
     1.5.3 Selective Laser Sintering (SLS) 7
     1.5.4 Material Jetting (MJ) 8
  1.6 Summary 9
  1.7 Chapter Problems 9
  References 10

2 3D Printing Applications Across Industry 11
  2.1 Objectives 11
  2.2 Overview 11
  2.3 Rapid Prototyping and Manufacturing 12
  2.4 Education and Academic Research 13
  2.5 Art, Fashion, and Jewelry 14
  2.6 Food and Nutrition 15
  2.7 Healthcare 16
  2.8 Housing 17
  2.9 Automotive 18
  2.10 Aerospace 19
  2.11 Summary 19
  2.12 Chapter Problems 20
  References 21

3   FDM 3DP Limitations 25
   3.1   Objectives 25
   3.2   Strength and Benefits of FDM 26
          3.2.1   Affordability 26
          3.2.2   Ease of Use 26
          3.2.3   Speed 27
          3.2.4   Flexibility 27
          3.2.5   Applications 27
          3.2.6   Scalability 28
   3.3   Weaknesses and Limitations of FDM 28
          3.3.1   Quality, Surface Finish, and Post-Processing 28
          3.3.2   Accuracy 28
          3.3.3   Longer Print Times 29
          3.3.4   Anisotropy and Strength 29
          3.3.5   Print Volume 30
          3.3.6   Consumer Safety 31
          3.3.7   Regular Maintenance 31
   3.4   Summary 32
   3.5   Chapter Problems 32
   References 33

4   FDM 3D Printing 35
   4.1   Objectives 35
   4.2   How FDM 3D Printing Works 35
   4.3   Variations in FDM 3D Printer Designs 36
   4.4   Getting Started with FDM 3D Printing 37
          4.4.1   Step 1: "Slice" the File 37
          4.4.2   Step 2: Load the Material 37
          4.4.3   Step 3: Start Printing 39
          4.4.4   Step 4: Post-Processing 40
   4.5   Common Cartesian Printer Anatomy 41
          4.5.1   A Basic Overview 41
          4.5.2   Axes of Movement 41
          4.5.3   Hot End Assembly 42
          4.5.4   Key Components 43
   4.6   Summary 43
   4.7   Chapter Problems 44
   References 44

5 Affordable Desktop 3D Printers 45
   5.1 Objectives 45
   5.2 Popular Brands 45
      5.2.1 Prusa 46
      5.2.2 Ultimaker 48
      5.2.3 Creality 50
      5.2.4 Lulzbot 50
      5.2.5 Flashforge 51
      5.2.6 Monoprice 52
   5.3 Getting Started with Prusa 53
      5.3.1 Why Prusas? 53
      5.3.2 Capabilities 53
      5.3.3 General Anatomy 54
      5.3.4 Prusa Hot End Assembly 56
   5.4 Summary 57
   5.5 Chapter Problems 57
   References 58

6 Common 3D Printing Materials 59
   6.1 Objectives 59
   6.2 PLA (Polylactic Acid) 59
   6.3 ABS (Acetonitrile Butadiene Styrene) 60
   6.4 PETG (Polyethylene Terephthalate Glycol) 60
   6.5 TPU (Thermoplastic Polyurethane) 61
   6.6 Specialty Filaments and Their Applications 61
   6.7 Summary 62
   6.8 Chapter Problems 62
   References 62

7 From 3D Object to Physical 3D Print: Slicing Software 63
   7.1 Objectives 63
   7.2 What are Slicers? What is G-Code? 63
   7.3 Where to Find 3D Models 64
   7.4 CAD Models 65
   7.5 Popular Slicers and How to Get Started 66
   7.6 Navigating Slicers 67
      7.6.1 Selecting Your Printer 67
      7.6.2 Importing a 3D Model 68
      7.6.3 Moving, Rotating, Scaling, and Arranging 3D Models 68

7.7   Print and Quality Settings Panels 70
        7.7.1   Quality Settings 70
        7.7.2   Profiles (Layer Height) 70
        7.7.3   Infill Density 71
        7.7.4   Supports 72
        7.7.5   Build Plate Adhesion 73
7.8   Viewing Your Selected Settings 73
7.9   Important Parameters to Consider 74
7.10   Slicer Tips and Tricks 75
7.11   Summary 76
7.12   Chapter Problems 76
References 77

8   Advanced Slicer Settings 79
    8.1   Objectives 79
    8.2   Overview 80
    8.3   Cura's Custom Settings Panel 80
            8.3.1   Quality 83
            8.3.2   Walls 84
            8.3.3   Top/Bottom 85
            8.3.4   Infill 86
            8.3.5   Material 87
            8.3.6   Speed 88
            8.3.7   Travel 89
            8.3.8   Cooling 91
            8.3.9   Support 91
            8.3.10 Build Plate Adhesion 94
            8.3.11 Special Modes 95
            8.3.12 Experimental 96
    8.4   Troubleshooting Common Issues 97
            8.4.1   Print Not Sticking to the Build Plate 98
            8.4.2   Print is Warping or Peeling Off the Build Plate 100
            8.4.3   Print is Stringing or Oozing 102
            8.4.4   Print Has Shifted During Printing 104
            8.4.5   Print is Under-Extruding or Not Extruding Enough
                    Material 106
            8.4.6   Print Has Support Scarring 107
            8.4.7   Print Has Undesired Wavy Lines or Ripples on Surfaces .. 109
            8.4.8   3D Printer is Clogged 110
            8.4.9   3D Printer Has Stopped Midway Through the Project 112
            8.4.10 3D Printer Filament is Grinding 113

	8.4.11	Extruder is Moving Erratically	115
	8.4.12	Nozzle is Scraping the Build Plate	117
8.5	Summary		119
8.6	Chapter Problems		120
References			120

9 Preparing to Print 121
- 9.1 Objectives 121
- 9.2 General Overview 121
  - 9.2.1 Double-Checking Your Slicer Settings 122
  - 9.2.2 Navigating the Menu 122
  - 9.2.3 Loading Filament 123
  - 9.2.4 Starting Your Print 124
  - 9.2.5 Unloading and Storing Filament 126
- 9.3 Best Practices 127
- 9.4 Summary 128
- 9.5 Chapter Problems 128
- References 128

10 'Steam-Building' Exercises 129
- 10.1 Objectives 129
- 10.2 Science Application: Digitizing Fossils 129
  - 10.2.1 A Paleontological Puzzle: Whale Diets Through Time 131
  - 10.2.2 Comparing Cats and Dogs 135
  - 10.2.3 The True Size of a Megalodon 137
- 10.3 Technology Application: Prototyping a Prosthetic Hand 139
  - 10.3.1 Downloading the Prosthetic Files 140
  - 10.3.2 3D Printing Instructions 141
  - 10.3.3 Additional Material 142
  - 10.3.4 Assembly 143
  - 10.3.5 What's Next? 143
- 10.4 Engineering Application: 3D Printed Arduino Robot 144
  - 10.4.1 3D Printing the Otto Robot .STL Files 145
  - 10.4.2 Assembling and Coding Otto 146
- 10.5 Art Application: Diy Musical Instruments 149
  - 10.5.1 Woodwinds 149
  - 10.5.2 Brass Instruments 151
  - 10.5.3 Percussion Instruments 153
  - 10.5.4 String Instruments 154

10.6   Math Application: Visualizing Math   155
        10.6.1   Geometry   156
        10.6.2   Calculus and Abstract Math   157
        10.6.3   Mathematical Art   158
10.7   Summary   161
10.8   Chapter Problems   161
References   162

Resources   165

Index   167

About the Author

Tyler Kerr, M.S. received a B.A. in Geoscience from Franklin & Marshall College in Pennsylvania in 2011, and an M.S. in Geology (Paleontology) from the University of Wyoming in 2017. His background in paleontology and interest in emergent technology led him to a career in 3D printing, 3D scanning, digital rendering, and digitizing museum collections. Today, Kerr manages the Innovation Wyrkshop makerspace, one of the largest academic makerspaces in the Mountain West. In addition to the Innovation Wyrkshop, he designed and currently oversees nine successfully operating makerspaces across Wyoming, making him a state-recognized authority on makerspace development and programming. For his work, Kerr was a recipient of the 2018 Laramie Young Professionals 20 under 40 award, the University of Wyoming's 2020 Employee of the Quarter award, and the 2021 Employee of the Year award. His academic interests include 3D printing, digitization, and developing creative, gamified, out-of-the-box nerdy ways to engage communities and teach complex topics in meaningful ways. With over 11 years of experience as an outreach coordinator and academic educator in Science, Technology, Engineering, Arts, and Math (STEAM), he aims to prove that everyone and anyone–even paleontological fossils like him—can be a maker.

3D Printing 101

1

1.1 Objectives

Objectives: After reading this chapter, the reader should have a firm foundation on the fundamentals of 3D printing, including:

- A basic overview of 3D printing today
- The history of 3D printing from the 40's to the present
- The major types of 3D printing available today
- An overview of fused deposition modeling (FDM) 3D printing
- An overview of stereolithography (SLA) resin 3D printing
- An overview of selective laser sintering (SLS) powder bed 3D printing
- An overview of the astounding capabilities of full-color material jetting (MJ) 3D printing.

1.2 Overview

T. Kerr, *3D Printing*, Synthesis Lectures on Digital Circuits & Systems,
https://doi.org/10.1007/978-3-031-19350-7_1

You may have seen 3D printing at school, in your local library or community center, or even on TV. It's an extraordinary technology used across an ever-growing list of industries, from movies and museums to mechanical engineering and the Met Gala. Today, you are very likely to see 3D printers in makerspaces, libraries, schools, workshops, dentist's offices, Michelin-starred restaurants, automotive assembly lines, hospitals, operating rooms, manufacturing facilities, construction sites, farms, jewelry shops, and even the space station!

This chapter presents a general overview of how 3D printing works and what types of projects you might be able to make with the more popular desktop 3D printers on the market today. In this chapter, we're going to review a wide variety of different types of 3D printing technologies but will focus primarily on the most popular and widely accessible 3D print technology on the market: fused deposition modeling (FDM) 3D printing. So when we discuss 3D printing in broad strokes later in this book, know that we're talking about FDM machines. If you're interested in integrating 3D printing into all sorts of personal or professional projects, this is the place to start.

1.3   What is 3D Printing?

In a very basic sense, you might think of FDM 3D printing like using a computer-controlled hot glue gun. Just like a hot glue gun, these 3D printers heat a cylinder of material and eject it out a hot nozzle. There are a few key distinctions that set the two technologies apart, however. Unlike actual hot glue guns, our "super-smart hot glue gun" 3D printer has a very tiny nozzle averaging only 400 microns (0.4 mm) in diameter. And instead of squeezing out a cylinder of hot glue, FDM 3D printers eject a thin string of molten "**thermoplastic**," which is simply a plastic designed to melt at a specific temperature (usually around 200–250 °C). Typically, two components of a 3D printer work together to build 3D shapes on the build plate. One component uses motors to grip and drive spools of thermoplastic **filament** forward through the nozzle at a regular rate. At the same time, another part of the 3D printer heats up the filament to a set temperature to allow it to ooze out of the nozzle at an established diameter and layer height. Working together, these components allow molten plastic to flow or "**extrude**" out of a tiny nozzle at a regular and predictable rate. Once extruded, fans quickly cool the molten plastic in place as the machine pilots the nozzle to different areas on the build plate. In such a way, a 3D printer can ultimately build a 3D shape out of nothing by printing thin lines of thermoplastic, one on top of another, layer-by-layer.

Thus, every 3D printer builds parts based on the same basic idea: a digital 3D model is turned into a physical 3D object by **adding or fusing** material one layer at a time. This is where the term "**additive manufacturing**" comes from. At its core, 3D printing is a very different method of producing parts compared to traditional subtractive manufacturing like CNC machining, laser cutting, or formative manufacturing like injection molding.

It's not *too big* a leap to suggest that a fourth industrial revolution driven by additive manufacturing is on the horizon. As traditional production-manufacturing shifts, we're on the cusp of an innovative desktop manufacturing revolution, thanks in part to technologies like 3D printing and rapid prototyping. 20 or 30 years ago, 3D printers may have only been accessible to advanced research labs. Now, those 3D printers are faster, easier to use, more affordable, and can fit in makerspaces, garages, libraries, workshops, homes, retail stores, and even space stations. Because this type of technology is so much more accessible, 3D printing can lower costs, save time and effort, and go far beyond the limits of traditional fabrication processes for product development. From concept models and functional prototypes to jigs, fixtures, or even end-use parts in manufacturing, 3D printing technologies offer versatile solutions for a wide variety of applications, and come in all sorts of shapes and sizes. And that means that 3D printers are revolutionizing how we can make almost anything.

1.4   History of 3D Printing

3D printing certainly feels a bit like a futuristic, science fiction technology, but it may be older than you think! While the desktop 3D printing revolution really started around 2009, the core concepts behind 3D printing were dreamt up in the 1940s, and the first 3D printer patents were filed in the 1970s and 1980s. If we rewind the clock to 1945, science fiction author Murray Leinster is believed to be the first to predict a futuristic "constructor" that feeds plastic into a moving arm and turns sketches and drawings into hardened physical models [1]. Jump forward to 1971, and Johannes F. Gottwald first patents a "Liquid Metal Recorder" that solidifies liquid metal into shapes according to the movement of the device [2]. By the mid-1980s, inventors such as Hideo Kodama, Bill Masters, Alain Le Méhauté, Olivier de Witte, Jean Claude André, and Chuck Hull all filed patents for a range of different primitive 3D printers. One of these inventors, Chuck Hull, was the first to release a commercial 3D printer in 1987, which he named the SLA-1 [3, 4]. One year later, in 1988, S. Scott Crump built the first fused deposition modeling printer, which was later commercialized by his company Stratasys in 1992 [5]. All of these early commercial 3D printers might have cost tens of thousands, if not hundreds of thousands of dollars today!

Fast forward to 2009, and the patents on fused deposition modeling 3D printers expire. From this, the very first low-cost, accessible desktop 3D printers were born. For the first time in history, 3D printers didn't cost hundreds of thousands of dollars. Instead, everyday innovators and hobbyists could buy them for under $2,000. Better still, as of 2022, you can buy a high-quality, entry-level fused deposition modeling 3D printer for less than $200! Let's dive into the most common varieties of 3D printers available on the commercial market today.

1.5   Common Categories of 3D Printing

Given their expensive and technical roots in research and industry, these machines might seem daunting at first. The reality is quite the opposite: 3D printers can be surprisingly easy to use, even if you don't consider yourself a particularly tech-savvy person. Gone are the days when you might have to write thousands of lines of code or program the machine's every X, Y, and Z movement by hand. Today, thanks to the open-source 3D printing community, you can drag-and-drop 3D models into 3D printing software, choose a handful of settings, load your material, and hit *Print*. The only thing you need to get started is a general handle on your project's end goals. For example, are you creating a hobby project that will sit on your desk? Will your project bear weight? Should it be waterproof or airtight? Does it need to look photorealistic? Will it need to have a high dimensional accuracy? These choices will influence what type of 3D printer you want to use.

With this in mind, the first speedbump newcomers run into with 3D printing technology is the types of 3D printers and materials that they should use. Acronyms such as FDM, EBM, SLA, MSLA, SLS, DMLS, MJ, and MJF can start to blend together pretty quickly. Rather than discuss all of these in detail, let's cover a very quick top-down overview. 3D printing is a relatively general term covering seven major additive manufacturing categories. Within those seven categories, there are around 18 different types of 3D printing technology. Every day, clever folks from around the world develop new techniques and technologies, so the list below will surely grow. At the time of this book's publication, 3D printing technologies [6] can be binned into the following categories:

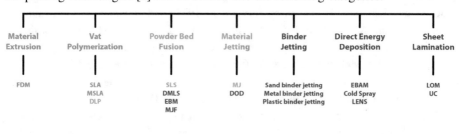

- **Material Extrusion**
  - *Fused Deposition Modeling (FDM)*
- **Vat Polymerization**
  - *Stereolithography (SLA)*
  - *Masked Stereolithography (MSLA)*
  - *Digital Light Processing (DLP)*
- **Powder Bed Fusion**
  - *Selective Laser Sintering (SLS)*
  - *Direct Metal Laser Sintering (DMLS)*

- *Electron Beam Melting (EBM)*
- *Multi Jet Fusion (MJF)*
- **Material Jetting**
  - *Material Jetting (MJ)*
  - *Drop on Demand (DOD)*
- **Binder Jetting**
  - *Sand Binder Jetting*
  - *Metal Binder Jetting*
  - *Plastic Binder Jetting*
- **Direct Energy Deposition**
  - *Electron Beam Additive Manufacturing (EBAM)*
  - *Cold Spray*
  - *Laser Engraved Net Shaping (LENS)*
- **Sheet Lamination**
  - *Laminated Object Manufacturing (LOM)*
  - *Ultrasonic Consolidation (UC).*

Rest assured, you won't need to learn all of these! Instead, we'll take a brief look at the four most common types of 3D printing that you're most likely to bump into in places like makerspaces, workshops, libraries, manufacturing facilities, or schools. These include fused deposition modeling (FDM), stereolithography (SLA), selective laser sintering (SLS), and material jetting (MJ). Following this, we'll focus on FDM 3D printing for the remainder of the book, as it is undoubtedly the most common 3D printing technology available.

1.5.1   Fused Deposition Modeling (FDM/FFF)

Fused deposition modeling, or "FDM" 3D printing (synonymous with another term: fused filament fabrication, or "FFF"), is a popular method of additive manufacturing where layers of melted thermoplastic are fed through and extruded out a small hot nozzle and fused together in a pattern to create a 3D object. The material is usually heated up just past the temperature that it begins to become viscous and melt, and then extruded in a pattern next to or on top of previous extruded layers, creating an object layer by layer (Fig. 1.1).

FDM is very accessible, user-friendly, and exceptionally well-suited for proof-of-concept prototypes, educational models, and quick and low-cost parts that might otherwise be machined more slowly. FDM can also print in a wide variety of strong, flexible, weather-resistant, hardy, or exotic materials. This includes materials like wood pulp to be sanded and stained, metal particulate to be polished and oxidized, chocolate to be molded into delicious geometric shapes, and even cement to build houses!

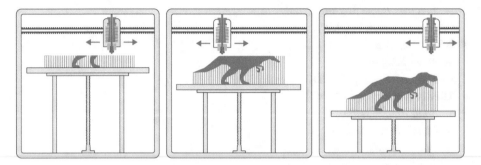

Fig. 1.1 The fused deposition modeling (FDM) process. A thermoplastic is ejected out of a hot nozzle and cooled in place by fans

1.5.2   Stereolithography (SLA)

Stereolithography, or "SLA" 3D printing, holds the interesting distinction of being the world's first commercial 3D printing technology, despite looking like something from a recent sci-fi film. Surprisingly, SLA was patented in 1987 by Chuck Hull, while in contrast, FDM was only commercialized in 1992.

An SLA printer uses special mirrors known as galvanometers that precisely angle and steer a laser beam across a container of liquid photopolymer **resin**. The pattern that the laser traces creates a hardened object layer by layer while the build plate and the cured object are pulled out of the uncured resin goo slowly (Fig. 1.2). It looks a bit like Han Solo coming out of carbonite.

SLA parts are known to have some of the highest resolution and accuracy, clearest details, and best surface finish of all desktop 3D printing technologies. Material manufacturers have created innovative SLA photopolymer resins with a huge range of different material properties tailor-made for engineering, industrial, medical, and research purposes.

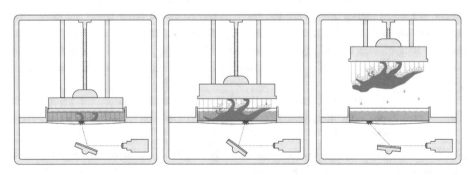

Fig. 1.2 The stereolithography (SLA) process. A liquid photopolymer is cured onto a build plate by a precise laser

SLA is excellent for highly detailed prototypes requiring tight tolerances and smooth surfaces, such as molds, patterns, and functional parts. Thus, SLA is used in a variety of
fields, including engineering, rapid prototyping, manufacturing, dentistry, jewelry, and
education.

1.5.3 Selective Laser Sintering (SLS)

Selective laser sintering, or "SLS," is another popular type of 3D printing, though it's a bit
less common than either FDM or SLA. This is largely because SLS machines use lasers
to fuse a bed of fine-grained powder, which usually means they can be a bit messier and
are often larger than traditional personal desktop 3D printers. They may also have unique
electrical, ventilation, personal protective equipment, and post-processing requirements
which means they're not *quite* as easily accessible to the average hobbyist. Still, they're
becoming increasingly popular, and thus worth noting briefly in this chapter.

If you can believe it, the SLS process was developed and patented in the mid-1980s
by an undergraduate student at the University of Texas named Carl Deckard, alongside
his academic adviser and mechanical engineering professor, Dr. Joe Beaman. It's worth
noting that Deckard and Beaman patented the SLS 3D printing process but certainly were
not the first to sinter objects. **Sintering**, the process of compacting and forming an object
using heat or pressure without melting the object, has been used for thousands of years.
Bricks, metals, ceramics, and jewelry all make use of the sintering process. If you've ever
packed a hard snowball by pressing loose snow together, you've practiced sintering.

Typically during the SLS process, tiny particles of powdered material like nylon plastic,
ceramic, or glass are fused together by heat from a high-power laser that traces the outline
of the object layer-by-layer (Fig. 1.3). As each thin layer of material is fused, the machine
lowers the print bed down a tiny amount and gently pushes another thin layer of material
across the top of the previous layer, a bit like a miniature bulldozer. This new layer is
subsequently sintered to the previous layer, and the whole process repeats. Gradually, the
3D printer forms a fused 3D object supported by the surrounding unsintered powder.

One of the most notable benefits of SLS over other 3D printing technologies is that
the powder bed filled with unsintered powder surrounding the 3D object can support the
object as it prints, which helps hold the entire thing together. As we'll learn later, FDM
and SLA 3D printing typically require overhanging parts (like arches, bridges, or bits that
stick out far from the model) to be supported by a latticework of material that is later
thrown away or recycled. While it's common for FDM and SLA models to require this
support, SLS parts use the surrounding material to do that job instead. This means that
despite being a bit messier, SLS parts require less post-processing or additional sanding
to tidy up an object once that object is 3D printed.

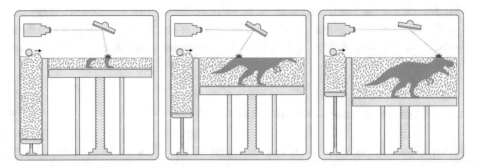

Fig. 1.3 The selective laser sintering (SLS) process. Thin layers of fine powdered material are selectively sintered (fused) by a laser

1.5.4   Material Jetting (MJ)

Did you know you can 3D print photorealistic, full-color, fully articulated objects? **Material Jetting,** or "MJ" (otherwise known as polyjet 3D printing), is arguably one of the most fascinating 3D printing processes commercially available today (Fig. 1.4). MJ works by spraying a line of rainbow-colored mist made up of incredibly tiny droplets of a liquid resin material onto a build plate, which is then cured in place by powerful ultraviolet (UV) lights. In a way, these 3D printers work a bit like a paper printer, though instead of mixing a single layer of cyan, magenta, yellow, and black (CYMK) *ink*, MJ printers spray extremely small droplets of CMYK photopolymer *resin*. After one layer has been laid down and cured by UV light, the build platform is lowered, more CMYK resin material is jetted out and cured, and gradually a 3D object forms.

Unlike other types of 3D printing that focus on a single point or area at a time and trace a path, MJ printers spray material in a line. This means that they typically zip back and forth, left and right fairly quickly. As a result, MJ printers are often a bit speedier

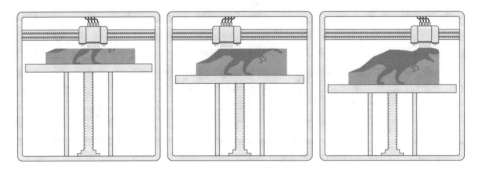

Fig. 1.4 The material jetting (MJ) process. Tiny droplets of liquid resin are jetted out of a print head and cured by UV light

than traditional FDM or SLA printers, although the cost of materials is typically a bit higher. Even though these printers are much more complex than conventional FDM or SLA printers, they're becoming more affordable and increasingly common in colleges, universities, and libraries. MJ is best known as one of the only types of 3D printing technology capable of printing full-color objects made of multiple materials simultaneously. Meaning that, yes, in theory you can 3D print an entire shoe.

1.6 Summary

This chapter provides a solid foundation on the fundamentals of 3D printing and additive manufacturing. We took a holistic approach, starting with a general overview of what 3D printing is, how it typically works, where it first originated, and what capabilities the technology has now. Next, we reviewed some of the most common types of 3D printing today and how to select the one that might suit your interests. The chapter detailed affordable desktop 3D printers, such as common fused deposition modeling (FDM) and stereolithography (SLA) machines. We then explored some extraordinary selective laser sintering (SLS) and material jetting (MJ) machines, which are becoming more and more affordable and accessible.

1.7 Chapter Problems

- How might you explain 3D printing to a general audience?
- How does additive manufacturing differ from subtractive manufacturing?
- What are some key differences between 3D printing in the 1970s and 1980s compared to today?
- What was the initial spark that made 3D printing more accessible to the public in the 2000s?
- How many categories of 3D printing currently exist? Which are the most common?
- Can you describe what FDM stands for? How does FDM 3D printing work?
- What role do "galvos" and lasers play in SLA 3D printing? How does SLA differ from FDM 3D printing?
- Can you describe sintering in your own words? How does it relate to SLS 3D printing?
- Why might you want to use an MJ 3D printer?
- What are some of the benefits of FDM, SLA, SLS, and MJ 3D printing?

References

1. Leinster, M. Things Pass By. Thrilling Wonder Stories (Stellar Publishing, 1945).
2. Gottwald, J. F. Liquid metal recorder. 5 (1971).
3. Charles Hull. National Inventors Hall of Fame https://www.invent.org/inductees/charles-hull.
4. Hickey, S. Chuck Hull: the father of 3D printing who shaped technology. The Guardian 1 (2014).
5. Crump, S. S. Apparatus and method for creating three-dimensional objects. 15 (1992).
6. Hubs Knowledge Base web page. https://www.hubs.com/knowledge-base/types-of-3d-printing/.

3D Printing Applications Across Industry

2

2.1 Objectives

Objectives: After reading this chapter, the reader should understand much more about how 3D printing can apply to a huge range of industries, not just engineering and advanced research fields. This chapter will include discussions and current-use case examples of 3D printing being used in incredibly diverse ways that include:

- Rapid prototyping and manufacturing
- Education and research
- Art, fashion, and jewelry
- Food and nutrition
- Bioprinting and healthcare
- Housing
- Automotive fields
- The aerospace industry.

2.2 Overview

3D printing is used in a huge range of industries. To highlight the sheer versatility of 3D printing, we should briefly explore some of the vast and varied 3D printing applications today. Suppose, for example, that you're looking to build the chassis for an Arduino robot. Maybe you want to design a beautiful geodesic dome out of chocolate instead, or construct a Martian habitat to keep space explorers safe. These are all exciting and innovative 3D

© The Author(s), under exclusive license to Springer Nature Switzerland AG 2022   11
T. Kerr, *3D Printing*, Synthesis Lectures on Digital Circuits & Systems,
https://doi.org/10.1007/978-3-031-19350-7_2

printing applications available today that hint at what the future of this technology may look like in the decades to come.

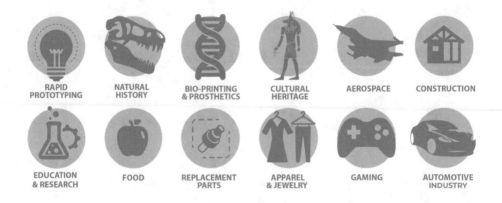

2.3   Rapid Prototyping and Manufacturing

With lower-cost desktop 3D printing more accessible and available to consumers today, anyone can make prototypes in an afternoon. The speed and efficiency at which 3D printers can produce custom parts make them a perfect tool for **rapid prototyping**. When we refer to rapid prototyping, we're referring to several techniques that enable anyone to quickly create functional models, parts, or assemblies from designs and sketches. For someone with a groundbreaking idea and a 3D printer, rapid prototyping can allow them to test out many different designs and quickly alter the size, shape, tolerances, and iterative variations of their design [1−4]. Conventional manufacturing often requires tooling, where manufacturers must create the tools to help build the unique product. Tools can include specialized equipment, molds, casts, dies, jigs, and custom parts. Often, tooling can take quite a bit of time, meaning that for innovators, it might take weeks or months between when they came up with an idea and when they finally hold that design in their hands. Rapid prototyping—specifically rapid tooling—allows innovators to bring better products to market faster than ever. Creative people with ideas can design and print physical prototypes in hours or days, not weeks or months. There are several key benefits that 3D printing and rapid prototyping bring to the manufacturing industry, including quicker lead times, more complex designs, unique one-off parts, custom on-demand tooling, and supply chain assistance [5].

3D printers can significantly reduce the lead time on custom parts or fixtures. Without any need for tooling, 3D printers can get straight to work producing parts. For manufacturers, this means time and money saved. Moreover, 3D printing can allow the production

Fig. 2.1 A custom ice core drill bit being 3D printed out of carbon fiber by the University of Wyoming Engineering Shop (https:// www.uwyo.edu/ceas/shop/)

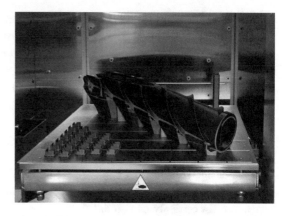

of parts that might be tough to produce in conventional manufacturing. Often in manufacturing, complex geometries that include 3D structures with undercuts or cavities would be considered complicated, time-intensive, costly, and challenging to create with traditional milling, turning, and casting. With 3D printing, there are few of these restrictions. 3D printing also produces considerably less material waste during the tooling and production process. One downside? 3D printing is better suited for small- and occasionally medium-batch production. Unless you have the funds to stack rows upon rows of 3D printers in a warehouse, the technology is not very well suited for large-scale production. At least not yet.

Even at a small production scale, 3D printers excel at producing unique one-off or on-demand parts (Fig. 2.1). If a manufacturer needs a particular part or tool made, they can do so in a number of hours instead of ordering and waiting for that custom part to be delivered.

Finally, 3D printing is excellent for producing spare parts or parts that can be digitally stockpiled to address future supply chain shortages. For example, suppose a part is no longer in production or perhaps too costly or difficult to procure. 3D printers can allow manufacturers to replace parts whenever they want instead of taking up physical inventory space. With digital files ready to be made in-house at a moment's notice, 3D printers can help to address and prevent supply chain shortages.

2.4   Education and Academic Research

Researchers and educators can use 3D printing to complement existing curricula or create custom parts, components, or tools to conduct new or niche research. For educators, you might think of 3D printers as a bit like Ms. Frizzle's Magic School Bus. They can scale microscopic life such as tardigrades up and place the 3D printed models in peoples' hands, or scale things down from the size of galaxies and do the same. They enable educators in rural and remote communities to download and produce exciting hands-on lesson plans for their students. 3D printers can also allow students of all ages and abilities to get direct and interactive engagement with objects they might never normally be able to see, such as rare museum collections or comparative studies. Biology students can 3D print complex cells, proteins, anatomical models, and animals. Engineering students can produce prototypes to assess the viability of designs, or print and test aerodynamic designs in wind tunnels. Culinary school students can print out detailed shapes and molds, stamps, and cutouts to cast or shape chocolate, dough, pastries, and gelatin. Entrepreneurial students can quickly create custom designs and prototypes and collect valuable market feedback. Graphic design students can explore new art mediums or have their artwork brought to life in 3D. History students can explore and examine rare artifacts that may be oceans away. The list certainly can go on!

By their nature, 3D printers excel at providing tangible teaching aides, 3D models, and physical prototypes to augment hands-on instruction [6, 7]. They can provide new constructionist learning opportunities through exploration [8]. For rural and remote communities, they provide access to tools, models, and digital files that may otherwise be inaccessible.

Within the academic research world, it should come as no surprise that 3D printing is particularly applicable. Researchers can prototype unique and custom-made devices, components, and instruments. They can experiment with custom-made materials and composites, or improve research design workflow and performance with new parts and methods. Furthermore, researchers can iterate designs and assess efficacy rapidly and at low-cost, conduct pre-research and pre-operative planning, and even replace costly or inaccessible parts that may retail for several orders of magnitude more money [9, 10].

2.5   Art, Fashion, and Jewelry

With the ability to 3D print sneakers, dresses, beachwear, and even jewelry, 3D printing is increasingly popular with artists and designers around the world. Today, there are 3D printers that can print on cloth, 3D printers that can print in metal or create castable metal molds in wax or plastic, and 3D printers that can print rubber and soft parts. Now, designers in the fashion industry can produce complex wearable geometry and accessories that traditionally may have taken much more time and effort to create [11, 12]. Better still, 3D printing allows designers to explore more efficient and even sustainable prototyping by producing only what is needed, and by recycling and reusing material [13, 14]. It's no wonder that today Nike and New Balance are using 3D printing to produce custom-fitted shoes for athletes, or that designers are harnessing 3D printers to make custom eyewear based on unique head shapes. With so many unique body shapes and sizes, and paired alongside technology such as 3D scanning, 3D printers seem tailor-made for the art, fashion, and jewelry world. Even more exciting, 3D scanning opens the door

Fig. 2.2 3D printed jewelry by Laramie, WY company Wyoming Brass Bone and Glass using 3D scans of fossils from the University of Wyoming Geological Museum fossil collections

for interdisciplinary science and art projects, such as 3D printed earrings created from museum displays (Fig. 2.2).

Within the broader art world, 3D printing goes hand-in-hand with creative expression. It provides a new medium to explore concepts of space, geometry, reconstruction, and reproducibility. Artists can go beyond sketches to visualize precisely what their artwork will look like in a physical space. Alongside 3D scanning, it can create accurate reconstructions of famous or historical artwork to emulate, reconstruct, and even modify. Furthermore, artists can produce identical 'photocopies' of their artwork to distribute globally with ease. Finally, 3D printing provides a low-cost and easy avenue to produce sculptures that may otherwise be difficult to produce using traditional methods.

2.6   Food and Nutrition

We're quickly approaching the world of Star Trek with food and bioprinting. Generally speaking, if you can make a material into a paste, squeeze it out of a large syringe, and harden it, you can 3D print it [15]. Chocolate, dough, candy, pancakes, hamburgers, and broccoli paste are all fair game. This type of 3D printing is increasingly used in restaurants, with exciting applications that have broad implications for reducing food waste and even producing food in space for astronauts and future planetary settlers. Today, companies are even exploring how to 3D print meat by cultivating stem cells to 'grow' fat and muscle within a 3D printed lattice! Considering that livestock farming contributes significantly to global warming, delicious lab-grown steaks have huge implications for the future of sustainable water, land, and livestock conservation [16–18].

Beyond its ability to produce creative designs and dishes, 3D printing brings several novel advantages to the culinary world. It allows restaurants, chefs, and foodies to create custom menu items with a calculated blend of vitamins, proteins, carbohydrates, and fatty acids catered directly to individual diners [19, 20]. Like we've now seen across many other fields, 3D printing also allows food production to happen in record time, and with less material waste. This ties directly into consistency and reproducibility. When you have a machine as precise as a 3D printer, it's easy to regulate ingredients to ensure that every diner receives the exact same nutritious, delicious, innovative dish.

2.7   Healthcare

If you can believe it, 3D printing functional human organs are only a heartbeat away. Scientists and engineers are working on methods to '**bioprint**' a carefully placed lattice-work of stem cells to produce new tissues and organs for humans. In fact, scientists and engineers at Tel Aviv University were recently able to 3D print the world's first beating human heart, complete with cells, blood vessels, ventricles, and chambers [21]! While the prototype heart is small—only the size of a rabbit's heart—it provides a clear glimpse at the technology hospitals might expect to have in the near future, and what might be possible on a human scale someday soon (Fig. 2.3). Equally exciting, being able to bio-print tissue and organs give surgeons endless opportunities to practice on exact replicas before actual surgeries [22] and opens the door for more compatible organ transplants. Bioprinted organs that are grown from each patient's own stem cells may be organs that our bodies are less likely to reject.

In addition to bioprinting organs and tissues, traditional 3D printing with plastics, rubbers, and metals plays a valuable role in healthcare. Like the fashion industry, 3D printers can be used to produce personalized and precision-made custom parts based on a patient's body size, age, sex, and lifestyle. This means that amputees can receive custom, comfortable, form-fitted prosthetics [23]. Likewise, dental patients can receive custom 3D

Fig. 2.3 A 3D printed heart within a support bath, from "3D Printing of Personalized Thick and Perfusable Cardiac Patches and Hearts" Noor et al. Advanced Science. © 2019 The Authors (CC BY 4.0) Scale bar: 0.5 cm

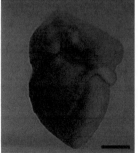

printed crowns from 3D scans of their teeth [24, 25], and patients can receive custom-fitted metal bone implants [26]. Hospitals can produce custom machine or device parts like ventilator valves or nasal swabs [27], and doctors can create teaching and practice models for education, research, and training.

2.8 Housing

Major construction companies and nonprofit organizations like Habitat for Humanity are betting big on 3D printing [28, 29]. Functionally almost no different than traditional Cartesian FDM 3D printers, there are currently giant 3D printers the size of a small plot of land that can 3D print entire houses out of unique concrete composites (Fig. 2.4). In such a way, construction companies such as Alquist 3D that work alongside Habitat for Humanity can 3D print a 1,200 square foot house in just over 24 h, and a small neighborhood in a week [30]. For reference, a typical Habitat for Humanity house takes four weeks to construct!

This type of 3D printing has profound implications for affordable housing and housing in developing nations. Construction company Alquist 3D calculates that 3D printing houses is not only faster, but cuts building costs by 15% per square foot. Ultimately, this means that 3D printing houses are cheaper, faster, safer, and more efficient—requiring less material with minimal waste, a smaller ecological footprint, and fewer on-site construction staff to complete.

Right now, companies around the world are experimenting with a number of different 3D printing technologies. The most common method of 3D printed houses is through enormous plot-sized 3D printers that use giant extruders to lay down a special blend of concrete layer by layer, while human teams on the ground work around the machine to install doors, windows, electrical components, plumbing systems, and insulation. Other architects and civil engineers like Enrico Dini are experimenting with sand 3D printing,

Fig. 2.4 The first owner-occupied 3D printed house, by additive manufacturing construction company Alquist 3D. Photo by Alquist 3D and Consociate Media

where a layer of fine-grained sand powder is laid down in large beds [31, 32]. Then, a liquid binding agent is selectively applied to the thin layer of sand, another layer of fine powder is recoated on top of the hardened powder, and the process repeats until a structure emerges. Finally, a type of wire arc 3D printing has been developed that uses robotic arms to weld together a latticework of stainless steel to create metal structures and—perhaps sometime soon—bridges [33].

2.9   Automotive

NBC's *Tonight Show* host Jay Leno is known not only as a comedian, but famously as an avid collector of vintage cars. Leno employs a small team to keep the almost 300 cars and motorcycles in his collection in peak driving condition, which is no small feat. Some of his cars are over 100 years old, which means that replacement parts are either scarce or entirely nonexistent. So what does his team do? They 3D scan, 3D print, and even cast the parts themselves! Leno's "Big Dog Garage" has several 3D printers in-house, and also contracts with large 3D printing companies such as 3D Systems to manufacture parts on more advanced SLS 3D printers. Leno noted in an interview that "with 3D printing, the automotive industry has changed more in the last ten years than it did throughout the entire last century" [34].

From Leno's use of 3D printing with his personal car collection, we know that 3D printing can allow hobbyists and enthusiasts to easily replace or upgrade parts. It should be no surprise that large manufacturers can do the same on larger scales. In fact, some major car companies are actively using 3D printers on the assembly line floor to produce decorative and functional plastic and metal parts, as well as jigs and fixtures to help expedite the assembly process. And that's not all. Automotive designers can produce and iterate upon prototypes quickly. Custom car shops can build and install unique parts in-house. Everything from body interior and trim to durable air and fluid management systems and metal components can be built by 3D printers [35, 36]. Believe it or not, some companies such as Desktop Metal have even designed ways of 3D printing photopolymer foam car seats [37].

Today, Aston Martin, Audi, BMW, Cadillac, Ford, GM, Hyundai, Porsche, Tesla, Volkswagen, and Volvo (likely among many others) are all utilizing 3D printing as part of their vehicle manufacturing process. 3D printing opens up the door to rapidly-produce custom high-precision parts for vehicles, while also helping ease and address supply chain issues to boot.

2.10 Aerospace

If you had to choose, would you wager that physically shipping up a set of tools to the International Space Station on the next Soyuz or SpaceX rocket or simply emailing the CAD files of those same tools is more efficient and cost-effective? When every pound of a resupply rocket's payload counts, you can bet that emailing up the 3D schematics for a wrench to the International Space Station is much easier than sending different tools up with every shipment! In fact, the aerospace industry (which includes research, technology, and manufacturing of materials and machines related to flight, aircraft, and spacecraft) is one of the most prolific adopters of additive manufacturing technologies [38–40]. Extraterrestrial zero-gravity FDM 3D printers, which are in their relative infancy, work with many of the same principles as terrestrial FDM printers but have a few more constraints [41]. Piloted rocket launches typically reach around three gs, or three times the force of gravity on Earth, and any payload (including 3D printers) must be able to withstand the forces of launch. When they reach their destination, these zero-gravity 3D printers must be sealed in an environmentally controlled container, and all moving parts of the machine must be secured well and unable to wobble around.

Today, NASA is investing quite a bit in zero-gravity 3D printers, food 3D printers for deep space missions, 3D printers to build parts for everything from the Perseverance Rover to orbiting satellites, and even construction 3D printers to build lunar and Martian bases. In fact, since 2014, astronauts aboard the International Space Station have had a zero-gravity 3D printer, which they use to 3D print CAD files transmitted up to them in order to build critical tools and equipment [42, 43]. You can freely download some of the parts and tools made on the ISS, including the first custom-made wrench, by visiting https://www.nasa3d.arc.nasa.gov/models.

Even more incredible, future explorers will be able to harness local materials to 3D print houses and equipment. In fact, one of the winning designs by AI SpaceFactory for NASA's Centennial Challenges competition proposed building 3D printed habitats for near and deep space exploration like Mars using Martian soil (Fig. 2.5) [44]. Rather than shipping building materials to Mars, these early Martian 3D printers will use Martian "regolith" (the layers of loose rock and soil on the Martian surface) to 3D print houses, landing pads, roads, and other structures for the next generation of space explorers.

2.11 Summary

3D printing has come a long way since the 1970s and '80s. In this chapter, we wanted to give you a glimpse into the vast and varied world of 3D printing today. We discussed how creative folks from all over the world and from all different disciplines apply 3D printing concepts to their work, research, and hobbies. 3D printers are changing the game and unlocking new opportunities in manufacturing through rapid prototyping, where parts can

Fig. 2.5 MARSHA, the
Martian habitat prototype
designed for NASA by AI
SpaceFactory. Image courtesy
of AI SpaceFactory

be made quickly, more affordably, and with less material waste, allowing manufacturers
and innovators to iterate and improve designs at a much faster and more efficient pace.
Educational applications are broad in scope and scale, providing instructional aides that
can augment hands-on education across almost all disciplines, and promote equal access
to critical resources. Researchers around the world are harnessing 3D printing to produce
affordable, custom-built parts, tools, and instruments, and are using 3D printers to ask
new questions and to plan and develop new innovative tests. 3D printing is a natural
fit within the art and fashion world, providing opportunities to explore new mediums,
produce unique or complex designs, emulate famous works, create custom clothes, and
even produce jewelry. We explored the ways in which food 3D printing is picking up
steam, where 3D printers are becoming increasingly common in restaurants, labs, and
businesses as a means not only to design creative new foods, but also to improve human
and animal nutrition and even address global environmental challenges. In healthcare,
we discussed how 3D printing is breaching the walls of science fiction through patient-
specific 3D printed organs, custom prosthetics, tools, and implants. In construction, 3D
printing is making big strides in affordable housing and housing in developing nations,
with the ability to 3D print an entire house in just over 24 h. Finally, we discussed how 3D
printers are being used in both the automotive and aerospace industries to create unique
parts and tools for all types of vehicles, improve the assembly line floor, and even build
the first Martian habitats.

2.12   Chapter Problems

- Can you effectively identify ways in which 3D printing is being applied to manufacturing? How is it changing the industry?

- In what ways is 3D printing being used in education and research? If you have a particular area of expertise, how might you use 3D printing in your own educational or research interests? What type of 3D printer would you use?
- How is 3D printing directly entwined in the art world? What exciting opportunities exist for custom or complex fashion and jewelry projects?
- How is the culinary and nutritional world being affected by 3D printing? Can you think of additional value that 3D printers bring to the world of food, diet, and health?
- Name a few ways in which 3D printing is affecting the healthcare world. Can you think of specific applications where 3D printing might help a family member of yours 5 to 10 years from now?
- What makes 3D printing houses so exciting? What implications does this have for affordable housing?
- Can you list a few of the ways in which 3D printing is changing the automotive and aerospace industries? Then, think 50 years into the future. How might you predict 3D printers will be used by 2072?

References

1. Stratasys, I. Rapid Prototyping. Stratasys web page https://www.stratasys.com/en/industries-and-applications/3d-printing-applications/rapid-prototyping/.
2. Formlabs. Bring Products to Market Faster With In-House Rapid Prototyping. Formlabs web page https://www.formlabs.com/applications/rapid-prototyping/.
3. Zortrax. Rapid Prototyping with 3D Printers. Zortrax web page https://www.zortrax.com/applications/rapid-prototyping/.
4. Vinodh, S., Sundararaj, G., Devadasan, S. R., Kuttalingam, D. & Rajanayagam, D. Agility through rapid prototyping technology in a manufacturing environment using a 3D printer. Journal of Manufacturing Technology Management 20, 1023–1041 (2009).
5. PWC. 3D Printing and the New Shape of Industrial Manufacturing. 1–22 https://www.forsythtech.edu/files/3d-printing-next_manufacturing-pwc.pdf (2014).
6. Ford, S. & Minshall, T. Invited review article: Where and how 3D printing is used in teaching and education. Addit Manuf 25, 131–150 (2019).
7. Assante, D., Cennamo, G. M. & Placidi, L. 3D Printing in Education: An European Perspective. in IEEE Global Engineering Education Conference, EDUCON vols 2020-April 1133–1138 (2020).
8. Pearson, H. A. & Dubé, A. K. 3D printing as an educational technology: theoretical perspectives, learning outcomes, and recommendations for practice. Educ Inf Technol (Dordr) 27, 3037–3064 (2022).
9. Silver, A. Five innovative ways to use 3D printing in the laboratory. Nature 565, 123–124 (2019).
10. Coakley, M. & Hurt, D. E. 3D Printing in the Laboratory: Maximize Time and Funds with Customized and Open-Source Labware. J Lab Autom 21, 489–495 (2016).
11. Sedhom, R. v. 3D Printing and its Effect on the Fashion Industry: It's More Than Just About Intellectual Property. Santa Clara Law Rev 55, 1–17 (2015).

12. Vanderploeg, A., Lee, S. E. & Mamp, M. The application of 3D printing technology in the fashion industry. International Journal of Fashion Design, Technology and Education 10, 170–179 (2017).

13. Lim, H.-W. & Cassidy, T. 3D Printing Technology Revolution in Future Sustainable Fashion. in 2014 International Textiles & Costume Culture Congress 25–26 (TD, 2014).

14. Pasricha, A. & Greeninger, R. Exploration of 3D printing to create zero-waste sustainable fashion notions and jewelry. Fashion and Textiles 5, (2018).

15. Natural Machines homepage. https://www.naturalmachines.com/.

16. Handral, H. K., Tay, S. H., Chan, W. W. & Choudhury, D. 3D Printing of Cultured Meat Products. Critical Reviews in Food Science and Nutrition vol. 62 272–281 Preprint at https://doi.org/10.1080/10408398.2020.1815172 (2022).

17. Jandyal, M., Malav, O. P., Mehta, N. & Chatli, M. K. 3D Printing of Meat: A New Frontier of Food from Download to Delicious: A Review. Int J Curr Microbiol Appl Sci 10, 2095–2111 (2021).

18. Ramachandraiah, K. Potential Development of Sustainable 3D-Printed Meat Analogues: A Review. Sustainability 13, 1–20 (2021).

19. Burke-Shyne, S., Gallegos, D. & Williams, T. 3D food printing: nutrition opportunities and challenges. British Food Journal 123, 649–663 (2021).

20. Sun, J., Zhou, W., Huang, D., Fuh, J. Y. H. & Hong, G. S. An Overview of 3D Printing Technologies for Food Fabrication. Food Bioproc Tech 8, 1605–1615 (2015).

21. Noor, N. et al. 3D Printing of Personalized Thick and Perfusable Cardiac Patches and Hearts. Advanced Science 6, (2019).

22. Carroll, D. 3D bioprinted heart provides new tool for surgeons. Carnegie Mellon University College of Engineering 1 (2020).

23. Manero, A. et al. Implementation of 3D Printing Technology in the Field of Prosthetics: Past, Present, and Future. Int J Environ Res Public Health 16, (2019).

24. Dawood, A., Marti, B. M., Sauret-Jackson, V. & Darwood, A. 3D printing in dentistry. Br Dent J 219, 521–529 (2015).

25. Formlabs. Digital Dentistry: 5 Ways 3D Printing has Redefined the Dental Industry. Formlabs Industry Insights 1 (2019).

26. Li, Z., Wang, Q. & Liu, G. A Review of 3D Printed Bone Implants. Micromachines vol. 13 Preprint at https://doi.org/10.3390/mi13040528 (2022).

27. Longhitano, G. A., Nunes, G. B., Candido, G. & da Silva, J. V. L. The role of 3D printing during COVID-19 pandemic: a review. Progress in Additive Manufacturing vol. 6 19–37 Preprint at https://doi.org/10.1007/s40964-020-00159-x (2021).

28. Cherner, J. Habitat for Humanity Debuts First Completed Home Constructed Via 3D Printer. Architectural Digest 1 (2022).

29. McCluskey, M. How the Company Behind TikTok's Viral 3D-Printed Houses Wants to Help Solve the Affordable Housing Crisis. Time 1 (2022).

30. Alquist3D homepage. https://www.alquist3d.com.

31. Sterling, B. Enrico Dini and His Architectural Sand Fab. Wired (2010).

32. Sher, D. One-to-one with Enrico Dini, the Italian who invented binder jetting for construction. 3D Printing Media Network https://www.3dprintingmedia.network/one-to-one-with-enrico-dini-the-italian-who-invented-binder-jetting-for-constructions/ (2019).

33. MX3D homepage. https://www.mx3d.com/.

34. Koslow, T. How 3D Printing Keeps Jay Leno's Massive Car Collection in Road Condition. 3D Printing Industry 1 (2016).

35. AMFG. 10 Exciting Examples of 3D Printing in the Automotive Industry in 2021. AMFG Knowledge Base 1 https://www.amfg.ai/2019/05/28/7-exciting-examples-of-3d-printing-in-the-automotive-industry/ (2019).

36. Formlabs. Road to the 3D Printed Car: 5 Ways 3D Printing is Changing the Automotive Industry. Formlabs Industry Insights 1 https://www.formlabs.com/blog/3d-printed-car-how-3d-printing-is-changing-the-automotive-industry/ (2019).

37. Sertogu, K. Desktop Metal Unveils New Expandable Foam 3D Printing Material, FreeFoam. 3D Printing Industry 1 (2022).

38. Altıparmak, S. C. & Xiao, B. A market assessment of additive manufacturing potential for the aerospace industry. J Manuf Process 68, 728–738 (2021).

39. Stratasys. Stratasys Aerospace webpage. https://www.stratasys.com/en/industries-and-applications/3d-printing-industries/aerospace/.

40. Lim, C. W. J., Le, K. Q., Lu, Q. & Wong, C. H. An overview of 3-D printing in the manufacturing, aerospace, and automotive industries. IEEE Potentials 35, 18–22 (2016).

41. Made in Space homepage. https://www.madeinspace.com/.

42. NASA. Solving the Challenges of Long Duration Space Flight with 3D Printing. NASA Mission Pages (2019).

43. NASA. Space Tools On Demand: 3D Printing in Zero G. (2014).

44. AI SpaceFactory homepage. https://www.aispacefactory.com/.

FDM 3DP Limitations

3

3.1   Objectives

Objectives: Each category of 3D printing has a long list of strengths and weaknesses that could fill this entire book, so for simplicity's sake, we'll focus here on the pros and cons of the technology you're most likely to bump into in your 3D printing journey: FDM 3D printing. Here, we'll cover:

- Strengths and benefits of FDM 3D printing, including:
  - Affordability
  - Ease of use
  - Speed
  - Flexibility
  - Applications
  - Scalability
- Weaknesses and limitations of FDM 3D printing, including:
  - Quality, surface finish, and post-processing
  - Accuracy
  - Longer print times
  - Anisotropy and strength
  - Print volume
  - Consumer safety
  - Regular maintenance.

T. Kerr, *3D Printing*, Synthesis Lectures on Digital Circuits & Systems, https://doi.org/10.1007/978-3-031-19350-7_3

3.2   Strength and Benefits of FDM

3.2.1   Affordability

Rewind the clock to the 1970s and '80s, and 3D printing would probably be considered quite cost-prohibitive for most people. Imagine trying to justify spending $10,000 or even $100,000 on a personal FDM 3D printer to a spouse or family member! Today, 3D printers are much more affordable and accessible, and FDM machines are among the least expensive 3D printers on the market. In fact, due to low part costs and simple designs, FDM printers are becoming continuously less expensive. Though some industrial machines still range into the hundreds of thousands of dollars, many introductory desktop 3D printers range in cost from $150 to $1000—significantly less expensive than their industrial counterparts. Likewise, filament is much easier to source and sometimes even make yourself. With so many companies producing 3D printing materials, it's becoming much cheaper too. A standard spool (usually 1 kg, or 1,000 g) of PLA filament averages around $15 to $25 and can last an individual user weeks or even months of regular use. This level of affordability of both the machines and the materials makes FDM 3D printing more accessible to hobbyists, educators, or small businesses that may not have the budget for large industrial machines.

Likewise, the nature of 3D printing usually means that labor cost and excess material waste is kept to a minimum. Since you can effectively load material, hit play, and walk away, there's less of a time investment required to run the equipment and produce prototypes. Similarly, by using only the material required to construct the part itself and very little else, there's much less material waste than in traditional manufacturing. All told, this makes 3D printing a very approachable technology that can be used for a huge range of applications, and within most budgets.

3.2.2   Ease of Use

3D printers today are considerably easier to use than the first 3D printers and CNC machines. Even though these machines all speak the same machine language, operators no longer need to manually program the movement and instructions for the machines themselves. Instead, a wide array of free, user-friendly 3D printing software simplifies the process. To get started, you only need to open a file, choose your preferred settings, load material into the printer, and hit *Print*. This means that 3D printing can be used by all types of people and all types of learning styles, and no longer requires significant training to operate. The simplicity of the software and the ease at which 3D printers can be used mean that they're perfect for beginners and even those that don't consider themselves particularly tech-savvy.

3.2.3   Speed

Another major advantage of FDM 3D printing is its speed relative to other traditional manufacturing methods. You can learn 3D printing in the morning and have that newest creation in your hands by the end of the day. Hobbyists, educators, researchers, and manufacturers can quickly design prototypes in-house, print them, make adjustments, and print again, all without investing significant time with the tooling and machining normally necessary in traditional manufacturing. Within industry, this usually means that production speed and cost are significantly cut, and that products can get to market faster.

3.2.4   Flexibility

3D printing offers much greater design flexibility compared to traditional manufacturing such as machining, molding, and casting. Your design doesn't meet specific tolerances? Forgot to add a key feature? Having trouble building complex or delicate parts? You can create and 3D print a revised design the next day. Moreover, with the ability to produce complex designs and intricate geometries, you can create single components. In contrast, traditional manufacturing might require several different parts later assembled together to make up the final part.

Likewise, FDM 3D printers offer a number of benefits when it comes to material flexibility. On many types of FDM 3D printers, you can use a wide variety of thermoplastic materials and specialty filaments out of the box, without having to invest in costly machine modifications. The wide selection of material types available and easily accessible allows FDM 3D printers to create parts out of everything from plastic to metal, carbon fiber, glow-in-the-dark material, and even wood-plastic composites.

3.2.5   Applications

As we covered in quite a bit of detail in Chap. 2, there are a vast number of applications for 3D printing today. Generally speaking, if there are creative and innovative folks in almost any discipline, there are clever applications for 3D printing within that industry. If you're jumping around in this book, we recommend that you take a look at Chap. 2 to learn all about how 3D printing is changing the game when it comes to rapid prototyping, manufacturing, education, research, art, fashion, food, healthcare, housing, and the automotive and aerospace industries—to name only a few.

3.2.6 Scalability

FDM printers can be built at almost any size, from desktop 3D printers to machines as large as a housing plot. The main constraint in the size of any FDM printer's build area is the range of movement of each gantry arm (called the X, Y, and Z axes). Simply put, if you make the gantry arms larger, the build area becomes larger too. Not many other 3D printer technologies can be scaled as easily with as few issues as FDM. We briefly touched on houses, entire neighborhoods, and even future Martian habitats being 3D printed using special cement and ground-up rock mixtures. Most of these are just particularly large material extrusion 3D printers, which work similarly to FDM desktop machines.

3.3   Weaknesses and Limitations of FDM

3.3.1   Quality, Surface Finish, and Post-Processing

One of the more noticeable disadvantages of FDM 3D printing is the part quality and detail. Because the material is extruded layer by layer at a specific layer height (resolution), very high-detail prints can be challenging or time-consuming to achieve. Most typical 3D printed parts come off the printer with a bit of a rough texture, and the layer lines of the print are often visible. This coarse surface finish may not be what you imagined for your project, but with some post-processing, you can still acquire a professional, finished look. Special epoxy sprays, chemical treatments [1], and sanding processes can make these parts look great, but take an extra bit of work (Fig. 3.1).

3.3.2   Accuracy

Many inexpensive 3D printers have looser tolerances compared to their industrial counterparts, meaning that the printed parts might have slightly different dimensions than the original design. This can be solved in post-processing through sanding and drilling, but

Fig. 3.1 An unfinished 3D print (left) compared to a chemically smoothed 3D print (right) by Polymaker (https://www.us.polymaker.com/)

might not be an ideal solution since any post-processing will increase the time and cost of producing the prototypes. More expensive machines tend to use more expensive parts and hardware, which means that they can usually ensure a greater level of precision and consistency and thus a tighter tolerance than introductory desktop FDM 3D printers. For most folks getting started with 3D printing, however, slight differences in tolerance are not too big of a concern. With some quick math and additional design iteration (i.e., scaling your project up by 5% to account for plastic shrinking), many introductory 3D printers can meet the needs of most basic projects. However, more robust industrial machines are likely the best route to explore for those requiring a very tight tolerance out of the gate.

3.3.3 Longer Print Times

Surprisingly, the speed of many desktop FDM 3D printers might be considered a drawback. 3D printing is quite speedy compared to conventional manufacturing, and can usually produce the majority of projects in less than a day. That said, it's still not the quickest process – at least for now. In 2017, some creative folks at MIT sought to address speed limitations, and prototyped an FDM 3D printer that can operate over ten times faster than regular desktop machines [2]!

To provide a bit of context, depending on the size and complexity of your project and the type of FDM 3D printer you're using, a typical 3D print made on a consumer-grade desktop FDM machine might take anywhere from 30 min to 30 h to complete. For example, a simple plum-sized object might take around 2 or 3 h, an apple-sized object might take roughly 5–6 h, and a cantaloupe-sized object might take 12 h. For some, these print times may be considered too slow.

3.3.4 Anisotropy and Strength

We've discussed how desktop FDM machines often don't have a perfectly smooth surface finish. Instead, there are visible layer lines that you can often feel if you run a finger across the print surface. These layer lines point to another limitation of FDM 3D printing. Printing a project layer by layer can create weak points at the spot where each layer is extruded on top of the preceding layer. Traditional injection molded parts, like LEGOs, might be considered "**isotropic**." In simple terms, these parts have very similar mechanical properties regardless of the direction you measure them. Broadly, solid injection molded parts would be uniformly stiff, strong, or stretchy if pulled, pushed, or smushed left to right, up and down, or forward and back. They have similar properties regardless of the direction measured.

In contrast, 3D printed parts from a typical FDM machine would be considered "**anisotropic**." This means that their physical properties vary across different directions.

Fig. 3.2 An example of anisotropy, where layers have begun to split and separate. © Prusa Research—Print quality guide—http://www.prusa3d. com

In FDM 3D printing, layers are typically laid down on top of one another along the Z-axis (up and down), and there are usually exceptionally tiny air gaps between each layer. Though uncommon, if the two layers don't adequately bond, there's a slight risk that the model might split apart and separate at this weak point (Fig. 3.2). Layer separation can make prints less sturdy and unsuitable for certain applications depending on which orientation the 3D model is printed. For example, you might not want to hang a heavy winter jacket on a 3D printed clothes hangar if that hangar was printed vertically with the hook sticking straight up.

3.3.5   Print Volume

Restrictions on the size and output of a 3D printer are additional factors to consider. FDM 3D printing isn't the best technology for significantly large or small projects. Even though there are countless different sizes of FDM 3D printers, most desktop FDM 3D printers have moderate build volumes, which means that creating parts larger than a watermelon or detailed parts smaller than a cherry can be challenging without the proper equipment. For example, Prusa i3 MK3S+3D printers, which we'll be working with later in this book, have a maximum build volume of approximately $9'' \times 8'' \times 8''$ ($25 \times 21 \times 21$ cm) [3]. Thus, larger objects would need to be made in multiple pieces on a typical 3D printer. Those who wish to print tiny, detailed parts might consider slowing down the 3D printer considerably to ensure adequate cooling between layers, as well as using a smaller diameter nozzle—shifting from a traditional 0.4 mm-wide nozzle to a 0.25 mm one.

Along these lines, FDM 3D printers excel at rapid prototyping and small-batch production but, as we discussed in Chap. 2, are not particularly well suited for medium- or large-batch production. Plenty of companies own warehouses filled with 3D printers and can mass produce parts at a large scale, but this isn't an option for most users with only one or two 3D printers at home. At larger scales, traditional manufacturing techniques such as injection molding and casting are better suited and more cost-effective options for higher volume production. However, with the relatively new introduction of conveyer belt 3D printers, which can print parts and roll them off the build platform until material runs out, this may change.

3.3.6 Consumer Safety

Typically, 3D printing projects are made of a few popular types of thermoplastic, such as ABS, PLA, PETG, and TPU. Of those, PLA is overwhelmingly the most popular due to its ease of use and affordability. This **PLA**, or "polylactic acid," is relatively biodegradable and (in the United States) usually derived from corn starch. In Asia, it's also made from tapioca roots and starch; in the rest of the world, it's sometimes derived from sugarcane. This means that PLA is an environmentally-friendly material. Aside from PLA, some thermoplastics, such as ABS, can release unpleasant and potentially toxic fumes. As such, setting up proper ventilation in the 3D printing area and discouraging hovering over 3D printers during operation is a good idea.

We've established that the process of FDM 3D printing can create small, microscopic gaps between layers. That means that FDM 3D printed parts are not likely to be airtight and watertight when fresh off the printer. If you pour water into an untreated FDM 3D printed container, you would see that container gradually start 'sweating' water over time through all of those tiny pore spaces. And unfortunately, where there are gaps and tiny pores, there can be bacteria. Thus, it's usually not the best idea to 3D print parts that you wish to eat or drink out of without first ensuring that the thermoplastic itself is food safe and that the part is fully sealed, especially if you wish to reuse it. Some thermoplastics such as ABS can withstand dishwasher temperatures and hold hot liquids. Others, such as PLA, cannot. If you wish to make 3D printed parts food safe, there are several sealants and post-processing measures you can explore that we won't detail here, but that a quick online search will provide.

3.3.7 Regular Maintenance

When a machine has more moving parts, or cheaper parts, it usually means that it may require more maintenance. Several large desktop 3D printing companies such as Prusa Research sell affordable 3D printers both as assembled machines and kits that you can build yourself, often at a discounted rate. Buying a kit is an excellent way to learn exactly how your 3D printer works, which can help when you need to do routine maintenance. The simple fact is that kits and inexpensive 3D printers with cheaper parts require more regular tune-ups to keep working. Many 3D printers like those from Prusa Research even use 3D printed parts in the design of their machines, and 3D parts can wear down over time. For those looking for 3D printers that work immediately out of the box and require less lifetime maintenance, you might consider spending a bit more.

3.4   Summary

This chapter explored some of the pros and cons of FDM 3D printers. We learned that FDM 3D printers, particularly desktop models, are among the most affordable and easy to use out of all 3D printing technologies. They are masters of rapid prototyping, and agile enough to allow designers to iterate and improve designs with a very quick turnaround time. They can be used for a vast variety of applications across science, technology, engineering, art, and math and can be scaled to almost any size with relative ease. Taken together, these benefits make FDM 3D printers among the most versatile and accessible 3D printers around.

In contrast, FDM 3D printers are not without their drawbacks. Typical desktop FDM 3D printers produce parts with a rougher surface finish than their SLA or MJ counterparts. These parts may look and feel coarse to the touch but can be sanded and chemically treated to make them smooth, if desired. Desktop FDM machines aren't necessarily known for their tight tolerances and high accuracy, though clever users can work to dial in these settings with time, or account for looser tolerances during initial design. FDM 3D printing, while fast, is still not quick and efficient as injection molding or casting at larger production scales. Instead, FDM 3D printing is best suited for rapid prototyping and smaller production runs. Due to the nature of how FDM 3D printers operate, parts can be a bit weak and porous. And last, we learned that there's always a trade-off: if you buy an inexpensive 3D printer, there's a good chance you'll need to learn how to maintain it yourself.

3.5   Chapter Problems

- What makes FDM 3D printers among the most affordable?
- How do 3D printers differ from traditional CNC technologies, despite speaking the same machine language?
- What are the pros and cons of FDM 3D printing when it comes to speed?
- What makes FDM 3D printing particularly flexible from a design standpoint? What about an applications standpoint?
- What size limitations exist for FDM machines?
- How might you address challenges with the surface quality of an FDM 3D printed part?

- What can you do if fasteners such as screws don't fit in the screw holes of a part you've designed and 3D printed?
- Can you think of any examples or projects where accuracy or anisotropy indicate that FDM 3D printing is not the best technology for that project?
- What options do you have if your project exceeds your printer's build volume? How might this affect production?
- What are the pros and cons of building an affordable FDM 3D printer from a kit?

References

1. Polymaker homepage. https://www.polymaker.com/.
2. Go, J. & Hart, A. J. Fast Desktop-Scale Extrusion Additive Manufacturing. Addit Manuf 18, 276–284 (2017).
3. Prusa homepage. https://www.prusa3d.com/.

FDM 3D Printing

4

4.1   Objectives

Objectives: This section will explore the FDM 3D printing process in more detail. Specifically, this chapter will review:

- How FDM 3D printing works
- Variations in FDM 3D printer design
- Getting started with Cartesian FDM 3D printers
- Common Cartesian FDM 3D printer anatomy.

By the end of the chapter, the reader should have a good grasp on how a typical FDM 3D printer operates from start to finish, and what role most of the core components of an FDM 3D printer play in the printing process.

4.2   How FDM 3D Printing Works

Fused deposition modeling 3D printers are incredibly versatile machines, capable of creating everything from board game pieces to prosthetic limbs and, as we discussed, even houses. As 3D printers become more common in businesses, schools, and homes, it's important to know how to use these machines for personal or professional projects. Read on to learn about how FDM 3D printing works, common FDM 3D printer anatomy, and tips and tricks for working with generic FDM 3D printers.

T. Kerr, *3D Printing*, Synthesis Lectures on Digital Circuits & Systems, https://doi.org/10.1007/978-3-031-19350-7_4

4.3   Variations in FDM 3D Printer Designs

Because they're inexpensive and simple to design, FDM printers come in a wide array of styles and shapes. For example, there are desktop 3D printers in schools used to create educational models and larger industrial FDM machines with built-in oven enclosures that can reach incredibly high temperatures to print materials to send into space. There are even food 3D printers that can print objects out of chocolate or broccoli paste in Michelin-starred restaurants [1]. Across these distinct variations of FDM 3D printers, the technology and design of the printer can differ.

Variations in FDM 3D printing usually revolve around the systems of movement for the printer's axes. The two most common variations in FDM printers are boxy "**Cartesian**" 3D printers (Fig. 4.1), which are named after the dimensional coordinate system (the X, Y, and Z-axis), and "**Delta**" 3D printers, which have circular build plates and an extruder suspended by three arms in a triangular configuration. More recently, there are robotic arm printers and "Polar" 3D printers. To keep things simple, the most important thing you need to know is that Cartesian printers are much more common than their counterparts and are the most likely type of FDM printers you'll encounter on your 3D printing journey. So how do we get started using a Cartesian FDM printer? First, let's walk through the steps.

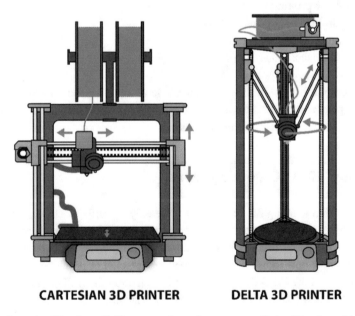

CARTESIAN 3D PRINTER   **DELTA 3D PRINTER**

Fig. 4.1 A Cartesian 3D printer (left) compared to a less-common Delta 3D printer (right)

4.4 Getting Started with FDM 3D Printing

4.4.1 Step 1: "Slice" the File

At its core, the FDM process always begins with a digital 3D model—essentially a blueprint of the physical object (Fig. 4.2). This model is sliced by 3D printing software (conveniently called "slicing software" or a "**slicer**") into thin, 2D layers. These layers and the instructions programmed alongside each layer (i.e. heat up to a specific temperature, print slower near the base of the model, squeeze out more plastic here, add support structures there) are then turned into a set of instructions in computer numerical control (CNC) machine language (a language called "**G-code**") for the printer to follow. You might consider this somewhat similar to the settings you choose when using a paper printer (i.e. print this document as a draft, at a high resolution, double-sided, in black and white, or print this in color). The difference is that the 3D printing slicer stacks those 2D slices up layer-by-layer to build a 3D model. Later in this book in Chaps. 7 and 8, we'll dive into slicers in more detail and highlight the quick and easy steps to get started with slicing software.

4.4.2 Step 2: Load the Material

Most FDM 3D printers can use a variety of plastics and other materials spooled up into rolls of wiry filament to create prints. Common FDM printing materials that you'll likely bump into as you start 3D printing include PLA and ABS, which we'll detail in Chap. 6.

In most FDM 3D printers, the material is loaded into the machine and extruded out the hot nozzle. In all desktop FDM cases, an extruder with a nozzle melts filament and pushes that molten filament out of the hot nozzle. FDM 3D printers have two core machine components: one part of the machine (the "**cold end**") uses drive gears to grip the filament, which steadily feeds the material toward the heat block and nozzle (the "**hot end**"). The hot end then regulates the material temperature in order to melt and extrude the plastic onto the bed at a predictable rate governed by the cold end. There are two main types of extruders that determine how the material is loaded and extruded: direct drive extruders and Bowden extruders (Fig. 4.3):

- Direct drive extruders (like those on the Prusa i3 MK3S+3D printers detailed later in this chapter [2]) seat the stepper motor drive gears directly next to the print head itself. This means that filament gets pushed directly into the hot end by the cold end, with relatively little space between the two. Direct drive extruders can be more reliable and easier to use, as filament is typically easier to load. They can also handle more unique types of filament. The biggest downside to direct drive extruders is that the print head

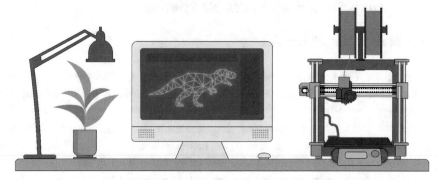

1. Download or design a 3D model

2. Slice the 3D model into layers

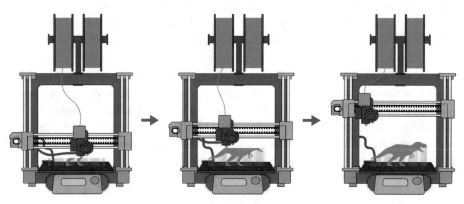

3. 3D print the sliced file layer-by-layer

Fig. 4.2 The three basic steps required to start 3D printing. Design or download a model, choose your settings and slice the file, and then 3D print it

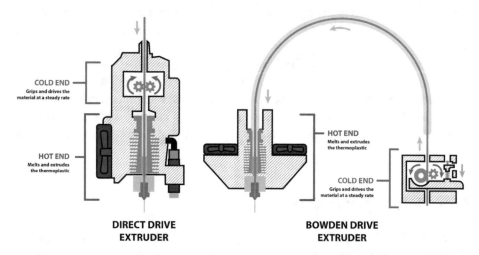

Fig. 4.3 Two most common types of extruders: a direct drive extruder (left) and a Bowden extruder (right)

must accommodate the weight of the cold end's stepper motor as the extruder travels back and forth, which makes direct drive extruders a bit slower than Bowden extruders.

- Bowden extruders (like those on Prusa MINI+3D printers) typically have the drive gears and cold end mounted on the frame of the 3D printer, usually some distance away from the print head and hot end. The cold end drives the filament through a long pliable plastic "**Bowden** tube" to the hot end and the extruder. A Bowden extruder setup means that the hot end doesn't have to support the weight of the cold end, which makes the machines lighter, quieter, and typically capable of producing higher resolution, more precise prints since it can start and stop more easily. The downside? Driving filament a greater distance requires beefier motors, which might drive up the cost of the 3D printer itself. Additionally, parts can be a bit stringier, with tiny spiderweb-like wisps of plastic that need to be cleaned off the final printed object.

4.4.3 Step 3: Start Printing

With your slicer settings selected, sent to the printer, and with the material loaded, it's time to start printing. Once the FDM printer receives the G-code instructions from the slicer, the hot end begins to heat up, and the printing starts. Recall from Chap. 3 that depending on the size and complexity of the 3D printed part and the type of printer; a print might take anywhere from 30 min to 30 h to complete. For example, a simple plum-sized object might take around 2.5 h, an apple-sized object might take roughly 5–6 h, and a cantaloupe-sized object might take 12 h. All told, the details of the object and the settings you've selected in the slicer all play a significant role in the time it will take to print.

To ensure everything is laying down and that material is extruding correctly, it's usually best to stick around and watch your project for a little bit to ensure there are no issues with the first layer of your print. Depending on the size of your project, this usually takes around 10–30 min. In a way, a 3D printed part is like a house: it can't stand without a sturdy, well-laid foundation. So, always make sure that the foundation is level and laid down properly.

4.4.4 Step 4: Post-Processing

Once your 3D printed part is finished, it might still require some touch-ups. Depending on your settings and the printer you used, 3D printed parts may not be ready to use immediately after the printing process stops. They may require some post-processing to achieve the desired level of surface finish. These steps take additional time and usually some manual effort to remove breakaway plastic supports or bed adhesion components such as brims, which we'll detail in Chap. 7.

We've established that FDM 3D printers work by depositing layer after layer of thermoplastic filament to create a 3D object. Each new layer is supported by the layers under it and quickly cooled by fans attached to the hot end once extruded from the nozzle. If your model has an overhanging part that is not supported by anything below it, the printer won't be able to print across that empty gap very easily, even with hot end fans cooling the plastic in place. You will need to turn on additional 3D printing support structures to ensure a successful print (Fig. 4.4). Once the print is finished, you can usually break off or dissolve away these **supports**, leaving only the original model. The downside of supports is that if they're too close to the model, not all the support material may break off, leaving some scarring on the bottom of your print that you might need to sand away.

On the other hand, **bed adhesion** (Fig. 4.5) is the ability of a 3D printed part to stick to the build plate while printing. When 3D prints have trouble sticking to the build plate, it's usually because 1) they're too small and the plastic hasn't had enough time to settle and cool between layers, 2) because the build plate didn't level correctly, or 3) because the plate isn't clean. When your project does not adhere to the bed properly, you can get curled, warped, and poor-quality results. Many times, the print will simply fail. 3D printers use varying methods to ensure that objects stick to the plate while printing. Most commonly, bed adhesion methods such as "**skirts**" or "**brims**" help prime the nozzle with melted filament, which increases the surface area of plastic touching the build plate. Any build plate adhesion would need to be peeled away and trimmed from the model after the print is finished, but we'll tackle build plate adhesion options in greater detail later in the book.

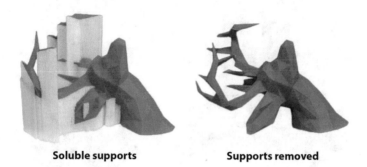

Soluble supports   **Supports removed**

Fig. 4.4 A 3D print with the soluble supports still attached (left) compared to a print with the soluble supports dissolved away (right)

Skirt   **Brim**   **Raft**

Fig. 4.5 Three types of bed adhesion: skirts, brims, and rafts. Screenshot from Ultimaker Cura (https://www.ultimaker.com) software

4.5   Common Cartesian Printer Anatomy

4.5.1   A Basic Overview

Here, we'll cover the basics of common Cartesian desktop FDM printers that you're likely to bump into on your journey toward 3D printing machine mastery. Between the names of the parts and their functions, it might seem challenging to keep it all straight. We'll keep it super simple here with a quick guide of the mechanical and electrical components in a generic FDM 3D printer (Fig. 4.6). Later, in Chap. 5, we'll deep dive into specific features found on Prusa i3 MK3S+3D printers.

4.5.2   Axes of Movement

1. The **x-axis** moves left to right, carrying the extruder along linear rods, rails, or reinforced rubber timing belts.
2. The **z-axis** moves either the extruder or the build platform up and down depending on your 3D printer. The z-axis is typically driven by lead screws, and sometimes

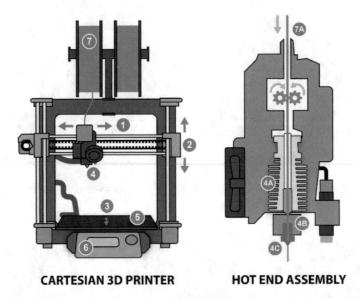

CARTESIAN 3D PRINTER   **HOT END ASSEMBLY**

Fig. 4.6 A Cartesian 3D printer (left) and a close-up view of the hot end extruder assembly (right)

guided by linear rods or rails. Lead screws (rather than rubber belts that usually control stepper movement motion along the X- and Y-axis) allow for finer motor control, which means more precise up and down controls. This, in turn, means finer resolution 3D printed parts.

3. The **y-axis** moves forward and backward. On Prusa printers, this moves the build plate (#5) forward and backward using a timing belt. On other printers, the y-axis may move the extruder forward and backward instead.

4.5.3 Hot End Assembly

4. The entire hot end assembly (in Fig. 4.6, a direct drive extruder) is made up of a few different components:

   A. The **heat sink** helps to dissipate the heat created by the heat block. It's separated by a small heat break, and prevents the intense 200–250 °C heat from creeping up the hot end too far, which can cause bad clogs. It's always best to turn off a printer sitting at temperature if you're not going to use it immediately. Otherwise, heat creep can occur, filament can start expanding where it shouldn't, and the hot end can clog.

   B. The **heat block** provides the heat to the nozzle through a heater cartridge and a thermistor.

C. The **nozzle** heats up and melts the filament with help from the heat block. The nozzle is screwed into the heater block. Nozzle diameters come in many sizes ranging from 0.25 mm to as wide as 1.2 mm or beyond, depending on the application. The most common nozzle size is 0.4 mm, which Prusa printers typically use. Always be cautious around nozzles. Depending on the material used, desktop FDM printer nozzles can reach upwards of 250–300 °C.

Note that Fig. 4.6 shows a direct drive extruder. This means that the hot end makes up the bottom half of the extruder assembly, and the cold end that drives filament makes up the top half.

4.5.4 Key Components

5. The **build plate** is where the material is extruded to form the 3D printed part. On an increasing number of FDM printers, these are not only heated (which makes 3D printed parts stick better) but also removable and flexible (which makes taking your part off the printer a breeze).
6. The user control interface is usually an LCD screen with buttons to navigate the menu. Some printers are controlled by nearby computers while others are operated entirely machine-side.
7. The **filament spool** is where the filament is placed during printing. This is usually found above, behind, or to the side of the printer. Filament (7A) is fed into the extruder, warmed up by the heat block (4B), and extruded out the nozzle (4C).

4.6 Summary

Now we're really starting to deep dive into the process of 3D printing! The goal of this chapter was to provide you with a robust background in 3D printing, with a particular focus on FDM 3D printing. We started with a top-down look at how FDM 3D printing works, and then hunkered down into variations among FDM 3D printers. Next, we looked at how to get started with FDM 3D printing and learned the key steps necessary to load a 3D model into a slicer. We explored several basic steps for loading material and how these methods may differ in direct drive and Bowden extruders. We then reviewed how to start a project and what types of post-processing work might be needed to perfect that project. Finally, we wrapped up by exploring common Cartesian 3D printer anatomy. We learned more about the axes of movement, how they move, how hot end assemblies work, and what components make up these hot ends. Finally, we wrapped up by discussing additional key components of most Cartesian 3D printers, including build plates, navigation menus, and material holders.

4.7   Chapter Problems

- Can you name some of the different variations of FDM 3D printers? Which is the most popular?
- What type of machine language do 3D printers use?
- Do you need to write and code G-code yourself?
- What is the purpose of a cold end?
- What is the purpose of a hot end?
- How do direct drive and Bowden extruders differ?
- In what way is 3D printing like building a house?
- What types of post-processing steps should you expect to have to do with 3D printers?
- In which direction does the X-axis move? The Y-axis? The Z-axis?
- Which axis typically carries the extruder left and right?
- Which axis typically moves the extruder or build platform forward and backward?
- Which axis controls the up and down movement of either the extruder or the build platform?
- What are the three primary components of a hot end?
- How does the nozzle become hot?
- How does the nozzle dissipate heat?

References

1. Natural Machines homepage. https://www.naturalmachines.com/.
2. Prusa homepage. https://www.prusa3d.com/.

Affordable Desktop 3D Printers

5

5.1 Objectives

Objectives: With a better handle on how the most common desktop 3D printers work and the purpose of most primary components of a printer, it's time to dive into popular brands and how to operate them. In this chapter, we'll discuss affordable and popular brands of desktop FDM 3D printers, which include:

- Prusa
- Ultimaker
- Creality
- LulzBot
- Flashforge
- Monoprice.

We'll then discuss how to get started using one of the most popular and affordable desktop 3D printers available: Prusa 3D printers. Finally, we'll end with a discussion on the specific capabilities and important anatomy of Prusa machines.

5.2 Popular Brands

Today, FDM 3D printers come in all sorts of shapes and sizes. If you're looking to learn how to integrate 3D printing into any personal or professional projects, there are a few key considerations you should factor in to ensure that you're selecting the correct machine for your interests. For example, what do you plan on printing? What sorts of materials or

© The Author(s), under exclusive license to Springer Nature Switzerland AG 2022
T. Kerr, *3D Printing*, Synthesis Lectures on Digital Circuits & Systems,
https://doi.org/10.1007/978-3-031-19350-7_5

colors do you want to print with? How large will your objects be? How high resolution must your projects be?

Below, we detail a few of the more popular, reliable, and well-known 3D printing brands on the market today. Then we'll hunker down for the rest of the chapter with one of 2022's best-in-class 3D printers: the Prusa i3 MK3S+.

5.2.1 Prusa

Prusa printers are considered an industry standard for affordable workhorse desktop 3D printers. Machines like the Prusa i3 MK3S+ and Prusa MINI+ are easy to use, reliable, and easy to fix (Fig. 5.1) [1]. Better yet, company founder and inventor Josef Prusa was a core developer of RepRap and makes a point of freely sharing out Prusa 3D files in case a 3D printer breaks down. Rather than buy new parts, Prusa makes it possible to just 3D print replacement parts for free! Today, the Prusa i3 design is considered one of the most popular FDM 3D printer designs on the market, and one adopted by hundreds of thousands of manufacturers and hobbyists worldwide, thanks in part to Josef Prusa's open-source and maker-focused approach. In fact, because of its award-winning record, the Prusa i3 MK3S+ is often considered a benchmark against which many other consumer-grade desktop FDM brands are compared.

The Prusa i3 MK3S+ is widely recognized as one of the best desktop FDM 3D printers available today, and one of the most frequently used 3D printers in the world. Originally released in 2019, the Prusa i3 MK3S and its successor, the i3 MK3S+, have won almost 20 distinct awards from prestigious technology companies and industry-leading magazines. At the time of writing this book, you can purchase i3 MK3S+ assembled machines for $1,099 or buy easy-to-assemble kits for $799. Notably, Prusa Research ensures that every design aspect of their 3D printers is available free and open source online, which makes modifying these printers and replacing parts simple and affordable. This open-source approach to innovation is a central driving force within the maker movement, which is part of the reason why these machines are so popular. Instead of competing with other brands or homemade Prusa variants, Prusa Research embraces the hobbyist modification movement and even tries to integrate some of the most popular open-source modifications into their commercial line to improve their machines further and address the needs of their consumers. The Prusa i3 MK3S+ boasts a $9.84 \times 8.3 \times 8.3$ inch ($25 \times 21 \times 21$ cm) build volume and can print 1.75 mm filament at a layer height ranging from 0.35 mm to 0.05 mm high. In addition, it has self-leveling build plates, flexible magnetic build plates, and a simple navigation system. These machines can utilize a very wide array of 3D printing thermoplastic filament, from PLA, PETG, ASA, ABS, PC, CPE, PVA, HIPS, PP, TPU, Nylon, and specialty filaments such as Woodfill. Considering the technology that comes standard in these machines, the freely accessible library parts and upgrades,

Fig. 5.1 Prusa MINI+ (top)
and the Prusa i3 MK3S+
(bottom). © Prusa
Research—http://www.pru
sa3d.com

and the active community and customer support resources, Prusa i3 MK3S+ 3D printers
are at the top of our recommendation list.

The Prusa MINI+ is one of the newer offerings from Prusa Research. These machines
are relatively affordable at the time of writing, retailing at $429. They're open-frame
printers that are small enough to sit on almost any desk or workbench, and quick to
assemble without much technical knowledge. The Prusa MINI+ features a $7 \times 7 \times 7$ inch
($18 \times 18 \times 18$ cm) build volume, and like most Prusa printers, comes with automatic bed
leveling, flexible and magnetic build plates that make removing parts a breeze as well as
an easy-to-use navigation screen. These machines use 1.75 mm filament and can print at
a layer height resolution ranging from 0.25 mm to 0.05 mm high. Moreover, like many
Prusas, the Prusa MINI+ is capable of tackling an extensive range of thermoplastics. The
Prusa MINI+ is a genuinely excellent and affordable entry-level 3D printer that should be
on the radar of all hobbyists, small businesses, crafters, makers, and engineers alike.

5.2.2 Ultimaker

If you're looking for a formidable workhorse machine with lots of bells and whistles, and are prepared to pay a bit more for some serious quality upgrades, then the Ultimaker S3 and S5 printers (Fig. 5.2) [2] should certainly be on your radar. For those looking for professional, dual-extrusion multi-material, and even soluble support printing capabilities, the Ultimaker printers might be high up on your wish list. These 3D printers are fast, sturdy, among the easiest to plug-and-play, require little maintenance, and can handle almost any challenging print you throw at them. Consider these machines if you have a larger budget and want a 3D printer that can handle tricky projects. There's a reason Ultimakers are consistently ranked as some of the best professional desktop 3D printers today.

Ultimaker machines retail significantly higher than most other desktop FDM machines, but with so many features available, they can be considered more advanced equipment intended for professional or industrial use. Ultimaker S3 machines cost around $4,450, which may deter hobbyists and small businesses on a budget. These are Cartesian, Bowden-drive, dual extrusion 3D printers with enclosed build plates, which makes them

Fig. 5.2 The Ultimaker S5
Pro. © Ultimaker (https://
www.ultimaker.com)

able to more efficiently regulate heat than open frame printers such as Prusa, LulzBot, or Creality machines. The Ultimaker S3 is the smallest of the Ultimaker fleet with printing enclosures, though a more affordable, compact Ultimaker 2+ is available to those who don't need a glass door to better regulate heat. The S3 features a $9 \times 7.4 \times 7.9$ inch ($23 \times 19 \times 20$ cm) build volume, and just like the larger S5, has two extruders and a range of perks. These include automatic bed leveling, swappable print cores, higher heat tolerances to print advanced materials, internet connectivity, and easy navigation controls. Unlike Prusas, Ultimaker machines use 2.85 mm filament and can print in two materials at once thanks to their dual extruders. For some users, this means printing in two colors. For others, it means two different materials, such as printing support latticework and bed adhesion features out of soluble PVA filament that dissolves in water. At the end of the day, the S3 can produce much smoother, higher-quality prints that can be printed at a layer height ranging from 0.2 to 0.02 mm high. All told, Ultimaker S3's are feature-packed, hands-off workhorses requiring little maintenance or oversight to consistently print high-quality projects.

Ultimaker S5 and S5 Pro machines share much in common with the smaller S3 3D printers. The key differences are size, material compatibility, and price tag. S5 printers are quite a bit wider than their counterparts, featuring two glass doors and a noticeably larger $13 \times 9.4 \times 11.8$ inch ($33 \times 24 \times 30$ cm) build volume. They also retail for $6,950 at the time of writing this book. Like their smaller cousins, S5's have dual extruders, a Bowden drive system, swappable print cores, automatic bed leveling, LCD screens, cameras, and internet connectivity. Similarly, they use 2.85 mm filament to print at resolutions that range from 0.2 to 0.02 mm high with standard 0.4 mm nozzles. Unlike the S3, however, Ultimaker S5's come with the added ability to use advanced print cores, such as the Ultimaker CC 0.4, which allows the machines to 3D print in high-strength materials like carbon fiber and even *metal* simply by switching print cores. Few, if any, affordable desktop 3D printers can do that.

The Ultimaker S5 Pros are effectively Ultimaker S5s with two critical key features added to enhance your projects even more. Perched above the Ultimaker S5 Pro is a smart airflow unit that helps regulate temperature within the unit, and underneath the S5 Pro is an automated, humidity-controlled material station that can house up to six spools of filament. In doing so, the S5 Pro allows you to print 24/7 for multiple days without having to monitor printers and swap out materials, which pushes this machine closer to the world of medium-batch production. Though pricey, the S5 and the S5 Pro are very reasonable options for advanced professional use compared to many other competitors at much higher price points. These machines are an excellent choice for hardy, easy-to-use, large-format 3D printers that require little to no oversight and offer some of the highest quality prints available.

5.2.3 Creality

Creality 3D printers [3] such as the Creality Ender-3 V2 and Ender-3 S1 Pro are excellent budget picks for those just getting started with 3D printing. Creality 3D printers can provide impressive print quality and reliability at a very reasonable price range, with the Ender-3 V2 retailing at around $250 and the Ender-3 S1 Pro retailing slightly higher at $479. Creality offers more than 25 different 3D printers that come in a wide range of build volume sizes, and are known for being hardy and reliable. More exciting still, Creality recently came out with one of the first commercial desktop 3D printing *conveyer belts* called the CR-30 that can print along the y-axis indefinitely. So if you ever wanted to 3D print a full-sized sword, consider Creality.

Creality's Ender-3 V2 is usually a top choice for those looking for sturdy, affordable, entry-level 3D printers. They share many physical similarities with Prusa 3D printers, including a similarly sized build volume of $8.6 \times 8.6 \times 9.8$ inches ($22 \times 22 \times 25$ cm). Like Prusas, the Ender-3 V2 works with 1.75 mm filament and a 0.4 mm nozzle. These machines don't quite have the layer height resolution range of Prusas but can still print between 0.4 mm and 0.1 mm high. The Ender-3 V2 has a textured glass build plate, a large color user interface LCD screen, and even under-bed tool storage. The machines can utilize PLA, TPU, and PETG filaments, which isn't quite as many materials as their slightly pricier Cartesian counterparts. Considering their price point and the size of the active Creality user community, the Ender-3 V2 is a reliable, affordable, excellent beginner's 3D printer that can handle 90% of the projects that a maker may wish to tackle in their first few years of 3D printing.

The Ender-3 S1 Pro can be considered a premium upgrade to the Ender-3 V2 line. While just under twice the price, the S1 Pro offers direct drive extrusion, self-leveling beds, a flexible magnetic build plate, an updated touch screen menu, lights to illuminate the work area, and a higher-grade metal hot end that allows the printer to reach up to 300 °C. This means that, unlike the Ender-3 V2, the S1 Pro can tackle PLA, ABS, PVA, TPU, PETG, and even wood-infused filament—to name a few. Like the Ender-3 V2, the S1 Pro boasts an $8.6 \times 8.6 \times 9.8$ inches ($22 \times 22 \times 25$ cm) build volume but can print a layer height resolution range between 0.05 and 0.4 mm high. While more expensive than the Ender-3 V2, the Ender-3 S1 Pro has sufficient bells and whistles to make it a popular favorite for beginner-to-intermediate users.

5.2.4 Lulzbot

LulzBot printers [4] are well-known as workhorses in the 3D printing world. These machines, whether the compact LulzBot TAZ Mini 2 or the configurable LulzBot TAZ SideKick, are considered easy to use and very reliable. Despite a higher price tag, 3D printers like the LulzBot TAZ 6, TAZ SideKick, and TAZ Workhorse have huge build volumes and are exceptionally dependable. So while hobbyist 3D printing enthusiasts might prize affordability over all else, educators, small businesses, and professional manufacturers looking for a solid and reliable 3D printer won't go wrong with LulzBot.

For a long time, the LulzBot TAZ 6 was LulzBot's flagship 3D printer. Even though the TAZ 6 retails for around $2,500, you get a considerable number of features for the money. These machines offer an impressive build volume of $11 \times 11 \times 9.8$ inches ($28 \times 28 \times 25$ cm). Like Prusa and Creality 3D printers, LulzBot machines are open and do not feature an out-of-the-box enclosure, though you can certainly purchase one. The TAZ 6 utilizes 2.85 mm filament and can print in PLA, PETG, ABS, PC, PVA, HIPS, and many different specialty filaments such as conductive and glow-in-the-dark thermoplastic. Depending on the nozzle installed, the TAZ 6 can print at a layer height ranging from 0.4 mm to 0.05 mm. For a reliable, sturdy 3D printer built in the United States, the Taz 6 is a great option.

However, there's a new exciting contender for top LulzBot 3D printer. The LulzBot SideKick is a configurable machine that starts at $795 and comes with a huge range of features you can modify and add to your machine before it ships. For example, you can choose the color of the machine itself, as well as the general size of the machine: either the TAZ SideKick 289, with a $6.375 \times 6.375 \times 7.125$ inch ($16.2 \times 16.2 \times 18$ cm) build volume, or the TAZ SideKick 747, which offers a $9 \times 9 \times 9.75$ inch ($22.9 \times 22.9 \times 24.8$ cm) build volume. In addition, you can select your preferred type of direct drive hot end, and can exchange yours for another easily. Those hot end tool heads come in various nozzle sizes and material types and can handle different diameter filaments. So, depending on your desired material use, you might select a hot end assembly that handles 1.75 mm filament with a brass nozzle or a hot end assembly that can tackle 2.85 mm filament with a hardened steel nozzle. You can also choose between glass build plates, flexible magnetic build plates, an LCD screen, a filament sensor, and a number of valuable accessories. For those who know exactly what types of projects they want to use a 3D printer for, and with a bit of knowledge of what machine features are needed to meet those goals, the configurable TAZ SideKick offers a lot of creative customization that's not often found outside of the grassroots modding community.

5.2.5 Flashforge

Entry-level Flashforge 3D printers [5] are prized for their plug-and-play simplicity and accessibility while also known for producing reliable, high-quality parts. As a result, machines like the Flashforge Finder are considered some of the best 3D printers for beginners, as well as young folks with little experience looking to dive into the world of 3D printing.

The Flashforge Finder is a compact direct drive 3D printer that retails for around $299. It offers a $5.5 \times 5.5 \times 5.5$ inch ($14 \times 14 \times 14$ cm) build volume and can print at a layer height resolution ranging from 0.5 mm to 0.1 mm. One of this machine's primary limiting factors is that it can only use PLA filament, which makes the Flashforge Finder best suited for younger makers or novices looking to learn the basics. In addition, the Finder features a non-heated glass build plate, which might be a bit tough for more complex 3D models to adhere to without help. That said, for the price point, the Finder is well worth the money—particularly for beginners.

For those looking to upgrade their Flashforge options, consider the Flashforge Creator Pro 2. This machine mirrors much of the Ultimaker's capacity, boasting a dual extruder capable of multiple materials or colors at once, but currently retails for an affordable $649. The Creator Pro 2 features direct drive extruders and a $7.9 \times 5.8 \times 5.9$ inch ($20 \times 14.8 \times 15$ cm) build volume—not the largest printable area we've seen, but fine for most projects. Notably, the Creator Pro 2 comes with IDEX ("independent double extrusion") technology, meaning that the machine can 3D print with the two extruders *simultaneously* instead of alternating back and forth between the two, as most dual extrusion printers do. This has some exciting implications, including speeding up 3D prints by almost half the time and printing in two materials at once. In addition, Flashforge Creator Pro 2 machines come with a heated BuildTak build plate and can print with PLA, ABS, PETG, HIPS, and PVA. Like the Flashforge Finder, the Creator Pro 2 can print at a layer height ranging from 0.4 mm to 0.1 mm. It does not come with automatic bed leveling, but the IDEX technology may make this a compelling option for those looking to explore dual extrusion 3D printing.

5.2.6 Monoprice

Monoprice printers [6] are among the best value and beginner-friendly 3D printers that you may be able to find. Home hobbyists and educators will get a lot of use out of Monoprice machines, such as the budget-friendly, preassembled Monoprice Voxel. These machines are easy to use, produce high-quality parts, and offer a fair number of features for the price.

At only $199, the Monoprice MP Cadet is an excellent, affordable machine for anyone looking for a basic introductory 3D printer to set up at home or in the classroom. It offers a $3.9 \times 4.1 \times 3.9$ inch ($10 \times 10.5 \times 10$ cm) build volume, and can print in 1.75 mm PLA material. Like other budget printers, it can 3D print at a layer height ranging from 0.5 mm to 0.1 mm high. Perhaps surprising for a less technical, budget-friendly printer, the Cadet is mostly plug-and-play and even offers Wi-Fi connectivity and automatic bed leveling. In short, the Cadet is an excellent option for safe-to-operate introductory 3D printing.

Fig. 5.3 Left to right, the Prusa MINI+, the Prusa i3 MK3S+, and the recently announced Prusa XL. © Prusa Research—http://www.prusa3d.com

5.3   Getting Started with Prusa

5.3.1   Why Prusas?

There are few brands with the impeccable reputation and enthusiastic fan base that Prague-based Prusa Research enjoys. Prusa printers (Fig. 5.3) are among the best desktop 3D printers on the market, some of the most reliable, workhorse machines, and surprisingly some of the most affordable, ranging from $400 to $1,000. In addition, they're accessible, easy to use, simple to repair and troubleshoot, and easy to modify, thanks in part to Prusa's willingness to make all their hardware, CAD files, and software freely available online. As a result, Prusa printers align well with the open-source 'maker' spirit ingrained in many aspects of makerspace culture today.

Since 2017, Prusa printers such as the Prusa i3 MK3S, Prusa i3 MK3S+, and the Prusa MINI+ have consistently won awards and dominated the charts for "best printer" and "Editor's Choice" as some of the highest-rated 3D printers on the market. As of 2022, the Prusa i3 MK3S+ is still considered the "Best 3D Printer," according to http://www. All3DP.com, PCMag, and the industry-leading Make: Magazine [7–9].

Simply put, these printers work, and they work very well.

5.3.2   Capabilities

The Original Prusa i3 MK3S+ 3D printer is one of the more recent designs to come out of Prusa Research in the last two years and features exceptionally high print qualities, with prints as fine as 0.05 mm (50 micron) layer heights. They have a large build volume of 9.84 × 8.3 × 8.3 inches and a direct drive extruder that can travel up to 200 mm/s, making them fairly speedy 3D printers. Prusa i3 MK3S+ machines are capable of automatic bed leveling, meaning there's no need to manually level the bed by hand. MK3S+ build plates are also heated, which does wonders to improve bed adhesion and guarantee successful prints. Better still, the build plates are magnetic, removable, and flexible, making retrieving finished 3D printed parts an easy process. Just gently flex the sheet and your print pops right off. Prusas use the more common 1.75 mm diameter filament, meaning that you can source inexpensive filament from reputable vendors fairly easily. And the Prusa i3 MK3S+ supports an impressive array of materials: PLA, ABS, PET, HIPS, TPU, PC, PP, Nylon, wood composites, metal composites, ASA, carbon fiber enhanced filaments, and many more.

Here, we'll take a deep dive into Prusa i3 MK3S+ anatomy (Fig. 5.4). Even if you have your eye on a different FDM printer, most FDM printers fundamentally operate the same way. By learning about the Prusa i3 MK3S+, you can still learn critical components that will help you in your 3D printing journey no matter what FDM desktop 3D printer you use. Let's check out the Prusa i3 MK3S+ in more detail.

5.3.3 General Anatomy

1. The extruder moves along the X-axis, left to right, using one rubber timing belt and two linear rods.
2. The extruder moves along the Z-axis, up and down, using two threaded rods and two linear guiding rods, one pair per side. Threaded rods, rather than belts, allow for better and more precise movement.
3. The heated flexible build plate moves along the Y-axis, forwards and backward, using a belt and two linear rods.
4. More recent Prusa printers have heated, flexible, removable magnetic build plates. These make it super easy to remove finished prints: just remove and gently bend the build plate and the print pops right off.
5. The hot end assembly is the part of a 3D printer that melts the filament and helps keep the machine at a consistent and accurate temperature to ensure successful prints. It's made up of several key components. Specific to Prusa printers, the hot end is made up of a direct drive extruder with two hobbed bolts (instead of a hobbed bolt and an idler bearing), a heat sink, a heat block, a brass 0.4 mm diameter nozzle, and two fans. One fan pointed at the heat sink to help it cool, another aimed at the extruded plastic.
6. One stepper motor controls the belt on the X-axis.
7. Two stepper motors control the up and down motion of the extruder along the Z-axis.
8. Threaded rods instead of belts along the Z-axis help to provide more precise up and down motion.
9. The On/Off switch is on the back of the power supply.
10. An LCD knob is used to rotate clockwise or counterclockwise and navigate the menu.
11. A reset button under the LCD knob allows you to click to cancel or turn off the heating options.

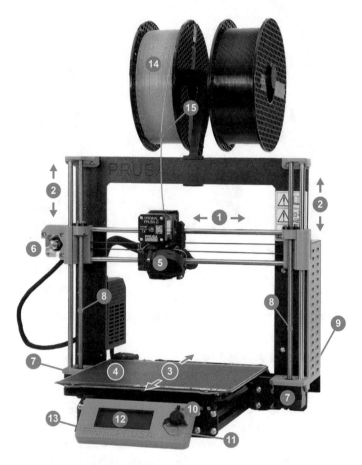

Fig. 5.4 Basic components of a Prusa i3 MK3S+ 3D printer. Annotated image modified from ©
Prusa Research—http://www.prusa3d.com

12. Navigate the LCD panel menu to load and unload the filament and import G-code
    projects from SD cards.
13. Prusa i3 MK3S+ machines use SD cards loaded into the LCD panel's left side.
14. Filament spools are placed on the filament holder.
15. Prusas use 1.75 mm diameter filament. It's always a good habit to unload and store
    filament in airtight bins. Atmospheric moisture in humid areas can seep into filament
    and cause prints to fail.

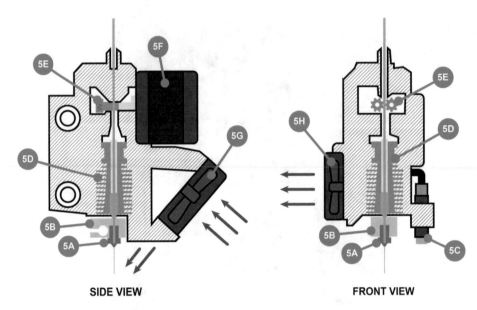

SIDE VIEW   **FRONT VIEW**

Fig. 5.5 Side view (left) and front view (right) of a Prusa i3 MK3S+ hot end

5.3.4 Prusa Hot End Assembly

Recall that the hot end assembly is the part of a 3D printer that melts the filament and helps keep the machine at a consistent and accurate temperature to ensure successful prints. It's made up of several key components. Specific to Prusa i3 MK3S+ printers, the hot end (Fig. 5.5) is made up of a direct drive extruder with two hobbed bolts, a heat sink, a heat block, a brass 0.4 mm diameter nozzle, and two fans (one pointed at the heat sink to help it cool, one aimed at the extruded plastic).

- 5A. The standard nozzle used by default in most Prusa printers is 0.4 mm in diameter.
- 5B. Above the nozzle is the heat block which, as its name implies, provides heat to the nozzle via a thermistor.
- 5C. The PINDA probe, short for "Prusa INDuction Autoleveling sensor," is a vital component to help calibrate the Prusa printers. It allows the printer to detect how far the nozzle is from the build plate.
- 5D. The heat sink helps to dissipate the high temperatures created by the heating block. If the printer is on at temperature for too long without filament moving through at a steady rate, heat creep can occur. Heat creep is the process of heat spreading irregularly throughout the hot end, which may create clogs and ruin the hot end. Therefore, it's a good habit to ensure that the extruder is not sitting at temperature when not in use.

- 5E. The hobbed bolts have teeth to grip and pull filament down through the direct drive extruder, past the heat sink and heat block, and down towards the nozzle. Newer Prusa models have two hobbed bolts to grip filament and provide tension, rather than a hobbed bolt to grip and an idler bearing to provide tension found in older Prusa machines (and more typical in most FDM printers). If filament is grinding, the hobbed bolt is often the culprit. Perhaps there's too much pressure? Too little? Maybe a filament tangle is preventing the filament from moving, so it's just grinding in place.
- 5F. The extruder stepper motor drives one of the hobbed bolts.
- 5G. A large fan aimed directly below the nozzle helps the cool down filament as it's extruded.
- 5H. Another fan aimed away from the heat sink helps pull heat from the hot end.

5.4 Summary

This chapter reviewed several popular 3D printing brands, including Prusa, Ultimaker, Creality, LulzBot, Flashforge, and Monoprice. We discussed what sorts of features make each unique, hoping to provide you with a comprehensive list of 3D printers to suit any project or budget. We then learned more about Prusa i3 MK3S+ 3D printers, which are considered to be among the best, most reliable 3D printers available today. We reviewed some of the features that make Prusa i3 MK3S+ best-in-class, and then covered the specific anatomy of the machine itself. We learned about how the printer moves and what major components do. We also explored the anatomy of the Prusa i3 MK3S+ hot end, with a special focus on how it regulates heat, drives filament, and probes the build plate. Our goal in this chapter was to provide sufficient information for anyone with a basic Cartesian 3D printer to locate and even troubleshoot challenges using a newfound understanding of the Prusa i3 MK3S+ primary operating procedure.

5.5 Chapter Problems

- Can you list some of the pros and cons of each type of printer? How do some differ?
- Are there any individual 3D printers or printer brands that best suit your project or idea?
- Is the Prusa i3 MK3S+ a Bowden drive or direct drive system? Can you recall the pros and cons of this type of extruder?
- On a Prusa i3 MK3S+, in which direction does the X-axis move? The Y-axis? The Z-axis?
- On a Prusa i3 MK3S+, which axis typically carries the extruder left and right?

- On a Prusa i3 MK3S+, which axis typically moves the build platform forward and backward?
- On a Prusa i3 MK3S+, which axis controls the up and down movement of either the extruder or the build platform?
- What diameter filament does a Prusa 3D printer use?
- Describe how the thermistor, heat block, nozzle, and heat sink are related.
- What does a PINDA probe do?
- What do hobbed bolts do?

References

1. Prusa homepage. https://www.prusa3d.com/.
2. Ultimaker homepage. https://www.ultimaker.com/.
3. Creality homepage. https://www.creality.com/.
4. Lulzbot homepage. https://www.lulzbot.com/.
5. Flashforge homepage. https://www.flashforge.com/.
6. Monoprice homepage. https://www.monoprice.com/category/3d-printing-&-hobbies/3d-printers-&-accessories/3d-printers.
7. Prusa Research. Original Prusa i3 MK3S+ web page. https://www.prusa3d.com/category/original-prusa-i3-mk3s/.
8. Hoffman, T. Original Prusa i3 MK3S Review. PCMag (2021).
9. Mensley, M. The Best 3D Printers of 2022 – Buyer's Guide. All3DP (2022).

Common 3D Printing Materials

6

6.1   Objectives

Objectives: We're almost ready to start 3D printing! This chapter will highlight many of the most common 3D printing thermoplastics [1−3] available today that you can use with FDM 3D printers such as the Prusa i3 MK3S+. Specifically, we'll review

- Polylactic Acid (PLA) thermoplastic, and why it's so popular
- Best use cases for ABS and PETG
- Additional use cases for niche thermoplastics such as flexible TPU
- The wide variety of fun and exciting exotic, composite, or specialty filaments

6.2   PLA (Polylactic Acid)

PLA is the most popular 3D printer filament type due to its ease of use, affordability, and dimensional accuracy. It's one of the easiest thermoplastics to work with and can be used on a wider variety of extruders and build plates due to its lower printing temperatures and resistance to warping and curling. Because of its popularity, PLA is available in an endless abundance of colors and styles, including exotic or specialty filaments such as conductive filament, glow-in-the-dark filament, and even filaments infused with metal or wood. Finally, PLA is a biodegradable thermoplastic usually made from corn starch, tapioca roots and starch, or sugarcane, meaning it's compostable in advanced commercial compost facilities. Thus, as far as 3D printing is concerned, PLA is one of the more environmentally friendly filaments you can use.

© The Author(s), under exclusive license to Springer Nature Switzerland AG 2022   59
T. Kerr, *3D Printing*, Synthesis Lectures on Digital Circuits & Systems,
https://doi.org/10.1007/978-3-031-19350-7_6

PLA is particularly popular in the makerspace community because of how easy it is to use. It's relatively reliable, resilient, largely odorless, doesn't necessarily require a heated build plate, and can be printed at lower temperatures than most other filaments. This makes PLA one of the most versatile materials available, and one that can be used with an enormous range of 3D printer brands. That said, PLA is not a one-size-fits-all solution for all projects. While it's relatively tough, it's still somewhat brittle and can sometimes fracture under stress. It can break down over time in sunlight or outdoor use, and is not ideal for projects that require a certain level of chemical resistivity. Finally, unlike materials like PETG, PLA is not food-safe without post-processing.

6.3   ABS (Acetonitrile Butadiene Styrene)

If you have toys, play music, own home appliances, watch TV, or use a computer, you've already interacted with ABS. This plastic is one of the most common materials used in injection molding today, meaning that you'll find ABS everywhere from home appliances, keyboards and computers, TVs, home appliances, toys, medical devices, plumbing and pipes, and vehicles. ABS is known to be strong, durable, heat-resistant, and has great mechanical properties. If anyone has ever stepped on a LEGO, they might be able to confirm how strong ABS can be!

ABS tends to melt at higher temperatures than materials such as PLA, which requires machines that can reach higher temperatures to melt and extrude the filament. Conversely, this means that ABS is less liable to melt or deform at lower temperatures. It's known to be more ductile, impact resistant, and less brittle compared to PLA, however, and can even be post-processed using chemicals such as acetone to create a smooth, glossy outer finish on a 3D print. ABS is also relatively chemically-resistant.

The downsides of ABS are simply that it's a bit more prone to warping and curling, and a bit more difficult to dial in perfect settings compared to PLA. Because ABS is prone to warping and requires a more regulated temperature, open-air 3D printers (such as Prusa, Creality, LulzBot, and Monoprice machines) may struggle with this plastic compared to enclosed printers such as Ultimakers or Flashforge. Similarly, ABS often requires a heated build plate to adhere well. Also worth noting, ABS often requires proper ventilation, as it can produce a pungent, potentially harmful odor.

6.4   PETG (Polyethylene Terephthalate Glycol)

Known for its flexibility, impact resistance, and as one of the most popular plastics in the world, PETG is another common type of filament you may want to use. You're likely to bump into PET and PETG with water bottles, food and medical packaging, and medical applications. Because PETG is nontoxic and FDA-approved, it's an excellent filament choice for food-related or medical projects, tools, and containers. Likewise, because it can be relatively weather-resistant, it's a great option for outdoor projects. PETG shares many of the same benefits as PLA while also mirroring the strength and durability of ABS. Unlike PLA, PETG is quite strong, and unlike ABS, PETG is less likely to warp and is mostly odorless while printing.

Worth noting: PETG can be challenging to dial in on a 3D printer and may stick firmly to the build plate after printing, potentially making it tough to remove. Because it's a bit sticky, PETG isn't ideal for 3D parts with significant bridging. You're also likely to have wispy 'spider web' stringing between parts of your project, which typically requires minor post-processing to remove and possibly sand down. Last, PETG is a bit more hygroscopic compared to materials such as PLA, which means that PETG filament spools should be kept in a dry environment so they don't absorb atmospheric moisture.

6.5   TPU (Thermoplastic Polyurethane)

If soft, rubbery, flexible 3D printing is an interest, consider using thermoplastic polyurethane (TPU). TPU is usually a composite blend of rubber and plastic, which makes this 3D printing material easier to bend, stretch, and flex. You can purchase different TPU filaments that vary in their elasticity and rigidity, with elastic properties similar to tires, wires and cables, through rubber bands. As a result, TPU is an excellent material for prototype drive belts, footwear, mobile tablet and phone cases, medical devices, handles and power tools, instrument panels, and even sporting goods. In addition, there's a high amount of customizability with TPU since flexibility is partially driven by the density of infill that you set, meaning that parts with little to no infill will be more flexible than parts with 20% or 50% infill.

Like PETG, TPU can be a bit sticky and thus can create lots of wispy stringing artifacts between parts that must be removed after the project finishes printing. It's also a bit more prone to create blobs and zits on the surface of the 3D print than other filaments. Therefore, TPU is usually best printed on direct drive extruders over Bowden drive extruders, though Bowden drive extruders such as Ultimakers can still easily handle TPU.

6.6   Specialty Filaments and Their Applications

Worth a brief mention is the wide range of specialty "exotic" filaments available. For adventurous makers, there are a huge range of different experimental filaments to play with: metal-infused plastic filament that can be polished and oxidized, wood-infused filament that can be printed with a grain and sanded or stained, conductive filament, carbon fiber filament, rainbow filament, marble filament, glow-in-the-dark filament, color-changing filament, shiny and reflective filament, coffee and beer filament made from food byproducts, and even soluble filament (PVA) that can be used to print support structures that can later be dissolved away in water.

6.7   Summary

In this chapter, we reviewed several of the most common 3D printing thermoplastics that you're likely to use as you start your 3D printing journey. First, we discussed PLA and why it's usually considered the 'default' filament for most makers. Next, we reviewed ABS and some of the benefits of using this durable, ductile material instead of PLA. We explored PETG and its use in food-safe or weather-safe applications. Then we briefly reviewed TPU and why you might consider 3D printing with flexible, rubber-like materials. Last, we discussed a growing list of specialty filaments that allow makers to 3D print with wood, metal, glow-in-the-dark filament, conductive materials, and much more.

6.8   Chapter Problems

- What are some of the benefits of PLA? What makes it the most used filament on the market?
- What are some of the challenges or limitations of using PLA?
- Why might you use ABS over PLA? What applications would ABS be better suited for?
- What are some of the challenges or limitations of using ABS?
- What types of applications is PETG used in the commercial world today? How might you use PETG filament?
- What are some of the challenges with using PETG?
- What can TPU be used for? Can you think of any applications in your life?
- What are some of the challenges with using TPU?
- There are many different types of specialty filaments available. Does any particular filament align well with a specific project you wish to explore?

References

1. Prusa Material Table web page. https://help.prusa3d.com/materials.
2. von Übel, M. 3D Printing Materials - The Ultimate Guide. All3DP (2021).
3. MatterHackers. Filament Guide web page. https://www.matterhackers.com/3d-printer-filament-compare.

From 3D Object to Physical 3D Print: Slicing Software

7

7.1 Objectives

Objectives: It's time to learn the software. This chapter will review how easy it is to get started with many different 3D printing software applications, even for those who don't consider themselves particularly tech-savvy. In particular, we'll review:

- What slicers and G-code are
- Where you might find and download millions of free 3D models
- A basic overview of computer-aided design (CAD) and how it factors into 3D printing
- How to navigate and use common slicer software to prepare a 3D model for printing
- Important introductory settings to consider for your first few 3D prints
- Tips and tricks for operating FDM printers, including what to do before, during, and after your project
- Common parameters and settings to consider, including support, infill, and build plate adhesion.

7.2 What are Slicers? What is G-Code?

What comes to mind when you hear the word "slicer?" Pizza, cake, and bagels might, but in the world of 3D printing, a **slicer** is a type of software that converts digital 3D models into printing instructions that a 3D printer reads to create an object. Many slicers are available online, but some of the most common ones include Ultimaker Cura, PrusaSlicer, OctoPrint, and Slic3r.

© The Author(s), under exclusive license to Springer Nature Switzerland AG 2022
T. Kerr, *3D Printing*, Synthesis Lectures on Digital Circuits & Systems,
https://doi.org/10.1007/978-3-031-19350-7_7

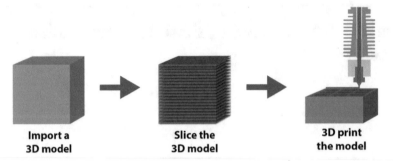

Fig. 7.1 The core concept behind slicing software, where a 3D model is split into thin layers and sent over to a 3D printer

These slicers do exactly what they describe—they slice. The software takes an uploaded 3D model and cuts that 3D model into horizontal stacked layers based on the settings selected (Fig. 7.1). The slicer then calculates how much material the printer will need to extrude to build the model, how thick each horizontal layer will be, and how long it will take to complete the project. This information is packaged into a **G-code** file and sent to the printer. G-code, in short, is a computer numerical control (CNC) programming language used in computer-aided manufacturing to control machines such as CNC lathes, mills, routers, and 3D printers. Slicer settings strongly influence the quality of your project, so it's important to select the best software and settings to get you the best quality 3D prints [1−4].

There are a large number of slicing software online and available to download, many of which are free. Even better, slicer software is as surprisingly easy to use as traditional 2D paper printing. Just load the model, select your desired settings, and hit *Print*. At their core, slicers are simply a way to get a 3D model from the computer to the 3D printer in a language that the 3D printing hardware can speak, and in a format (sliced, stacked layers) that the 3D printer can produce layer-by-layer.

This section will cover the basic slicer settings that every beginner should know. Later, we'll cover the steps necessary to start your 3D printing project, such as loading, unloading, and storing filament, navigating the Prusa menu, and monitoring the initial layers of the project.

7.3   Where to Find 3D Models

One of the great aspects of 3D printing is that you don't necessarily have to create models yourself. However, if you need to create a custom tool or part for a specific prototype or design project, it certainly pays to have a basic knowledge of CAD modeling to design that 3D object from scratch. For introductory 3D printing, however, this may not be the case. For those just getting started, there are many online repositories with *millions* of 3D files for you to access, download, and even modify. Consider visiting:

- https://thingiverse.com/
- https://grabcad.com/library
- https://www.turbosquid.com/
- https://sketchfab.com/
- https://www.cgtrader.com/
- https://www.printables.com/
- https://www.myminifactory.com/
- https://3dprint.nih.gov/
- https://www.yeggi.com/
- https://3d.si.edu/

7.4 CAD Models

If you want to build custom-made models from scratch, particularly 3D models that must have specific dimensions or tolerances, computer-aided design (CAD) software is often necessary. CAD software makes it possible to build objects to the specifications you need, meaning that CAD is incredibly useful for making repairs and building custom parts or fittings. Often, as is the case with other 3D file types shared online, you may be able to use existing libraries such as GrabCAD or Onshape to find an approximate design that you can then modify. We won't dive into CAD too much in this book other than highlighting its use in customizing some of the 'STEAM-building' exercises in Chap. 10 using free CAD files available online. However, if you're interested in learning more about CAD, we recommend exploring the following resources:

- Beginner, introductory use:
  - Tinkercad: https://www.tinkercad.com/learn
- Intermediate use:
  - Onshape: https://learn.onshape.com/
  - Autodesk Fusion 360: https://help.autodesk.com/view/fusion360/ENU/courses/
  - Autodesk Inventor: https://knowledge.autodesk.com/support/inventor/learn-explore/

- Advanced, professional use:
  - SolidWorks: https://my.solidworks.com/training/catalog/list/
  - AutoCAD: https://knowledge.autodesk.com/support/autocad/learn

7.5   Popular Slicers and How to Get Started

For the rest of the chapter, we will use **Ultimaker Cura** software, a powerful and very popular free slicer (Fig. 7.2). There are a large number of slicing softwares out there, many of which are free, so you might wish to use a different program. PrusaSlicer, the slicer developed specifically around Prusa machines, is another fine alternative. However, this book will focus primarily on Cura since it's more applicable across a wider array of 3D printer makes and models [1]. Whatever your slicer of choice may be, the principle settings and concepts you should understand remain the same.

Cura is one of the most popular and accessible slicers, and a favorite with beginners and advanced users alike. Cura may seem very simple at first glance, with only four settings to change, but hidden underneath the simple and accessible Cura exterior, however, is a treasure trove of hundreds of advanced settings to customize if you need and want to. It allows advanced users the ability to change almost any setting needed during the printing process while remaining simple enough on the front-end for most basic use. Users who don't care for all the bells and whistles and just need a part made only need to focus on

Fig. 7.2 Ultimaker Cura is one of the most popular free slicers available today. Image from Ultimaker (https://ultimaker.com)

four settings to start 3D printing. To download Cura for free, visit https://ultimaker.com/software/ultimaker-cura.

7.6 Navigating Slicers

7.6.1 Selecting Your Printer

When you start Cura for the first time, you'll need to select a printer (Fig. 7.3). Later, if you want to set up an additional printer, you can navigate to *Settings > Printer* and add or manage printers. Because Cura is so widely compatible across a wider array of 3D printer manufacturers, the selection list for printers is long and comprehensive and will likely have your machine settings available. Most printers listed in the Cura printer selection panel already come preprogrammed with their specific features and dimensions, which saves you quite a bit of work. There's no need to program your build volume, startup instructions, or preferred G-code language. Instead, Cura does all of that for you.

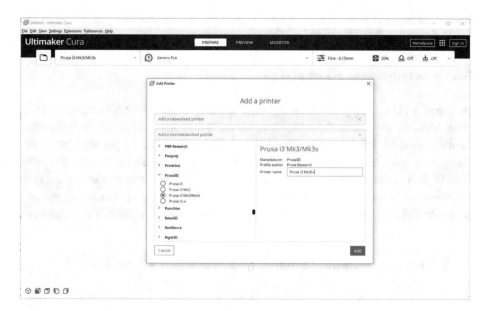

Fig. 7.3 Adding a specific 3D printer make and model in Ultimaker Cura. Screenshot from Ultimaker Cura (https://ultimaker.com) software

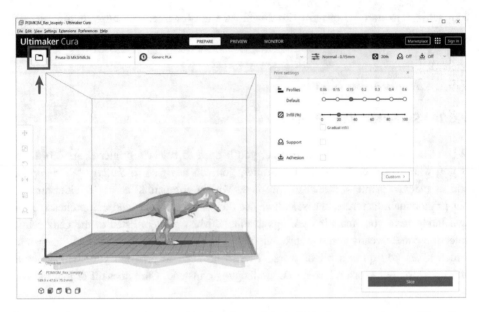

Fig. 7.4 To import a 3D model into Ultimaker Cura, click the folder icon, or navigate to *File > Open File(s)*. Screenshot from Ultimaker Cura (https://ultimaker.com) software

7.6.2 Importing a 3D Model

To import a 3D model, you can simply drag and drop a 3D model onto the digital build plate. You can also select the folder icon on the left or navigate to *File > Open File(s)* from the top menu (Fig. 7.4). From here, you'll need to select the proper file type, as most slicers have a limited number of preferred 3D file types they can work with. STL, OBJ, or 3MF files are most typical, though increasingly, CAD files can be directly imported. Note here that Cura works in metric, so if you've created a CAD file or 3D model in American software, you may have to do a bit of quick math to convert it to millimeters.

7.6.3 Moving, Rotating, Scaling, and Arranging 3D Models

When the model loads in the build area, it may be too small or too big for your project end goals. You may want to rotate the model at an angle, move it around the build plate to make room, or lay a specific planar surface flat on the build plate. Cura's lefthand toolbar allows you to adjust your 3D model in several ways (Fig. 7.5):

- The *Move* button (T) allows you to click and drag your 3D model anywhere on the build plate.

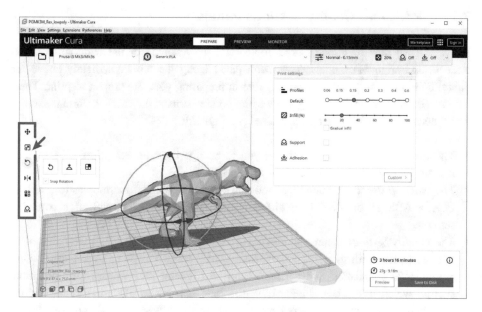

Fig. 7.5 The toolbar on the left allows you to move, scale, and rotate 3D models in Ultimaker Cura. Screenshot from Ultimaker Cura (https://ultimaker.com) software

- The *Scale* button (S) allows you to scale up or down the model by hand, metric measurement, or a percentage of the original file's size.
- The *Rotate* button (R) allows you to rotate the model freely, snap it to certain degrees of rotation, and even select specific faces of the model that you wish to lay flat against the build plate.

In addition to standard orientation buttons, you can click a specific 3D model with the right mouse button to access additional tools. For example, if you want to 3D print duplicates of an object, you can click the model with the right mouse button and select *Multiply Selected*. Cura will duplicate and automatically reposition all models. Models that appear in color indicate that there is enough space to print those 3D models. Otherwise, gray and yellow striped models indicate that there isn't enough space to fit that model on the build plate. Gray and yellow-striped models will be ignored and will not print.

Lastly, you might want to import a large number of files into Cura at once. While you can manually position these around the build plate yourself, you don't have to. To easily and automatically arrange all models on the build plate, use the right mouse button to select the build plate itself and select *Arrange All Models* (Control + R). From here, you can make minor position changes as desired. Once you've arranged and oriented your 3D files the way you'd like them to print, it's time to select some basic settings.

7.7   Print and Quality Settings Panels

We've learned how to utilize Cura's toolbar on the left side of the window to move, arrange, and align 3D models. The menu panel along the top immediately above the digital build plate plays another critical role in preparing your 3D project to print. Three buttons (*Printer*, *Material*, and *Quality Settings*) are perhaps the most important settings you'll need to consider to get your desired print quality.

- The *Printer* dropdown menu allows you to change between 3D printers, as well as manage and add new printers.
- The *Material* dropdown menu lets you quickly select different thermoplastics, such as PLA or ABS, all of which have different material properties that will affect available settings.
- The *Quality Settings* menu includes layer height resolution profiles (extra fine, fine, normal, draft), infill, supports, and adhesion. There are two menu options under Quality Settings: *Recommended* and *Custom*. Cura will default to *Recommended* on initial startup, which is the best choice if you're first learning to 3D print. Here, options are whittled down only to the four most crucial settings: layer height, infill, build plate adhesion, and basic support structures. *Custom* print options are available for advanced users who may wish to adjust hundreds of customizable settings, which we detail in Chap. 8. You can control almost every aspect of your 3D printing project using these custom settings, from initial first layer fan speed through custom-designed supports. For now, we'll stick to the *Recommended* panel.

7.7.1   Quality Settings

The print quality setting's menu options are the most important print parameter settings to consider. This section will review each setting in detail and discuss why you might want to change each setting.

7.7.2   Profiles (Layer Height)

The **layer height** of a project dictates the **resolution** of your print and specifies the height of each new layer of filament extruded by the nozzle. Parts printed with smaller layers will create more detailed, higher-resolution prints with a smoother surface (Fig. 7.6). In the case of these higher-resolution prints, it may be difficult to see the individual filament layers. Consequently, the object may appear close in quality to a smooth injection molded

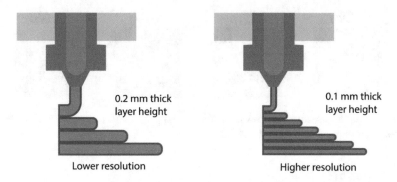

Fig. 7.6 The resolution of a 3D print is determined by the size (thickness) of each layer on top of the other

part. The trade-off, however, is the time that it would take to print a part with a small layer height: since more layers make up the object, print time may double or even triple.

If you wish to print something at a standard resolution, you might consider selecting a 0.15, 0.2, or even 0.3 mm resolution. These parts will have thicker layers and a rougher surface finish where individual layers may be clearly visible. Parts printed at these normal or draft resolutions will print faster than 0.1 mm or 0.06 mm resolution prints. This type of standard or low-resolution printing is fantastic for rapid prototyping projects where capturing extra fine details may not be as important. For most projects, a standard 0.15 or 0.2 mm layer height is just fine.

7.7.3 Infill Density

Infill refers to the density of the latticework of support structure inside a 3D printed object (Fig. 7.7). This infill provides structural rigidity and allows the model to print complex internal structures, roofs, and walls of the 3D model. If an object is printed with 100% infill, it will be completely solid on the inside. A part printed with 0% infill will be fully hollow and often thin and delicate. There's a trade-off: the higher the infill density, the stronger and heavier the object will be, but the more time and material it will take to print. If you're creating an item for display, we might recommend 5–10% infill. If you need something functional and sturdy, or something potentially undergoing significant stress or strain, 50–70% infill may be more appropriate. Only in specific situations do we find it necessary to print anything at 100% or 0% infill.

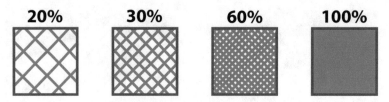

Fig. 7.7 Infill is the structural latticework inside the 3D print itself. Denser infill leads to stronger parts, but at the expense of time and material

7.7.4   Supports

Recall from earlier in the book that **supports** are external structures that surround your model and help to prop up any parts of the 3D model that have too steep an overhang or too long a horizontal bridge. Think of these supports like the latticework of crossbeams, rebar, and girders that would hold up a bridge under construction. You can't build a bridge out into horizontal empty space without supports holding up those new parts of the structure. The same rule applies to 3D prints. Once a 3D print is done, the plastic supports can be snapped off or dissolved away in water, depending on your 3D printer and its capabilities. Typically, you'll need to toggle on supports for any object with overhanging or bridging components extending beyond a 45° angle. Without support, the parts might droop, sag, or even fall over, which may ruin the project. Supports on Prusa printers are most often printed in the same material as the model itself. These supports are designed as "breakaway" supports—that is, they are designed to break off easily once the project is finished.

For some helpful background, there are two main types of 3D printing supports: those made from the same material as the object being printed, and those made from a different material (sometimes even water-soluble material). Each has its own advantages and disadvantages, but both types can be situationally useful depending on the purpose and desired quality of your project.

Soluble supports are typically made from materials like PVA, which can be dissolved in water or other solvents after the print is complete. PVA supports are ideal for complex prints with intricate details, as there is no need to remove the support structures manually. However, soluble supports can be more expensive than non-soluble ones, and they may not be strong enough for certain applications.

Non-soluble supports, on the other hand, must be removed manually after the print is complete. This can be a time-consuming process, but it allows for greater flexibility in terms of material choice and support structure design. Non-soluble supports are also stronger than soluble ones, making them better suited for large or heavy objects.

How do you know whether your design needs supports? Cura will tell you in the *Prepare* view by highlighting the model's overhanging and bridging features in red. Another

Fig. 7.8 The Y-H-T rule. Parts with bridges or overhangs (H, T) typically require support, while parts that gradually build upon each other (Y) may not

handy trick is the common **"Y-H-T" rule** (Fig. 7.8). If you were to 3D print the letters Y, H, and T, each might require different support settings:

- "Y"-shaped features with gradually sloping parts may be acceptable to print without supports toggled on because there is likely sufficient material overlap from layer to layer to keep the model from failing. However, slopes steeper than 45 to 60 degrees may still need support.
- "H"-shaped features with a horizontal bridging structure should have supports underneath that bridge to prevent failure or a messy, stringy final product.
- "T"-shaped features similarly would need supports under overhanging parts on both branches of the "T" to avoid failure.

7.7.5   Build Plate Adhesion

Whether or not you toggle on build plate adhesion and the build plate adhesion type you choose may affect the likelihood of your print succeeding or failing. Build plate adhesion affects how well your 3D model sticks to the build plate. If parts are too small or have minimal surface area touching the build plate, there may be an increased chance for your print to become detached from the build plate. When in doubt, it doesn't hurt to have adhesion toggled on. By default, Cura will enable a **"brim"** for the build plate adhesion type if you toggle on this setting.

Like a brim of a hat, 3D printed brims are the lines around the bottom of an object. Brims help increase the surface area of the first layer touching the build plate and also help to keep the corners of your model from peeling or warping up (Fig. 7.9). Brims can also stabilize or link together delicate parts of an object that may be isolated from the rest of the project.

7.8   Viewing Your Selected Settings

Until now, we've primarily worked in Cura's *Prepare* view. This allows us to change many settings on the fly without having to inspect the model after every change. With your layer height, infill, support, and adhesion settings now selected, it's time to take a

Fig. 7.9 Brims are a common type of bed adhesion, and are used if projects are having trouble adhering to the build plate

look at Cura's other view panels. First, select the *Slice* button in the bottom right corner to commit your settings. Next, we'll navigate to *Preview* at the top to review our model with the settings implemented. Worth noting, in Cura, there are three ways to view the model, each of which has a different utility that you might find useful. The default view when importing files, *Prepare* gives you a good idea of how the digital model appears and allows you to change important project settings. In contrast, the *Preview* tab we've just selected provides two view types: *Layer view* and *X-ray* view.

Within the *Preview* tab, the *X-ray view* provides a translucent view of your 3D model, which may be helpful if you need to inspect a 3D file for missing or glitchy geometry. We rarely use this view type, as there are other programs to inspect and repair files. *Layer view*, however, is one of the more important Cura windows that you should utilize.

Once you've selected settings and are ready to print, you should always hop into *Preview > View Type: Layer view* and inspect your model to ensure that it will print correctly. Changing the color scheme to *Line Type* helps visualize support and brim structures. A slider on the right side of the UI will allow you to scroll up and down layers of the model (Fig. 7.10). As you get more comfortable in Cura, *Layer view* can be critical to ensure that your model is oriented on the build plate correctly. It's a good habit to check the *Layer view* and scroll down to inspect the first few layers of your project before starting to print. Remember, 3D prints need a good foundation. If insufficient model or support material is touching the build plate, the print may not succeed.

7.9   Important Parameters to Consider

Remember, when first starting out, there are only four essential parameters to consider. All of these influence time, amount of material, and overall quality of your project, so before you print, think critically about the following criteria:

- What layer height resolution would you like on this project? Do you need a higher resolution, like 0.1 mm or even 0.06 mm, or is it okay to print at a standard resolution of 0.15–0.2 mm? Most projects are perfectly fine printed at 0.15 mm or 0.2 mm.

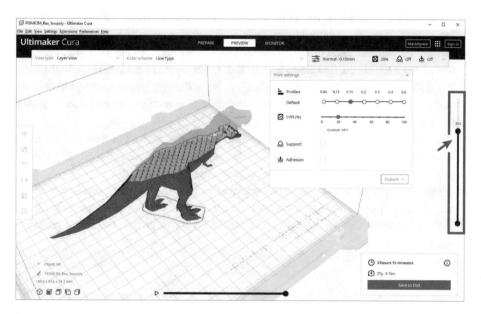

Fig. 7.10 Use *Preview* mode's *Layer view* to scroll up and down and assess your sliced model. This is a good method to ensure the success of a project. Screenshot from Ultimaker Cura (https://ultima ker.com) software

- What percent infill does your project require? Most 3D printed parts that simply sit on display are fine with a 5–10% infill density. Of course, stronger parts require denser infill, but we rarely find ourselves needing anything beyond a 50% infill density.
- Does your project require supports? Projects with significant overhangs, bridges, or projects that generally have T- or H-shapes or parts will likely require supports, whereas projects with gradual angles (Y-shapes) that have a sufficient foundation from previous layers do not.
- Does your project require some type of build plate adhesion? Projects with many small parts arranged near each other on the build plate will likely need a brim, while larger objects with plenty of surface area touching the build plate may not.

7.10   Slicer Tips and Tricks

- Cura has hundreds of different settings and a plethora of profiles available for dozens of different printer brands. Because of this, it may feel overwhelming when looking at the complete list of settings. Cura defaults to *Recommended* settings for users, and for the most part, this is the only list of settings you should need when you're first learning. *Custom* settings can come later once you're comfortable with the basics.

- Always get into the habit of double-checking your settings. As we'll discuss, once your model has been sliced, be sure to look at it in *Layer view* to ensure nothing looks amiss. It's good to double-check that your project is sitting flat on the build plate and that nothing is floating in midair. A poor foundation is one of the most common reasons for failed prints.
- Remember to save your model as G-code for the printer, as well as a project file to edit later if need be. This will allow you to fix any errors that present themselves during printing without reconstructing the project settings from scratch.
- When troubleshooting, like any good engineer or scientist, it's good practice to change only one setting at a time and observe the results. This is the best way to track how each incremental change affects your print. Don't try to change too many settings at once.

7.11   Summary

In this chapter, we dove into slicing software, with a particular focus on one of the most popular and powerful slicers available today: Ultimaker Cura. First, we learned about G-code, and where to find and even make 3D models to 3D print. We then explored how to navigate within Cura, starting with the printer selection menu and then the import options to open 3D models in the slicing software. Next, we learned how to move, scale, rotate, duplicate, and arrange 3D model files in the software to position our files in the desired orientation that our project may call for. We explored the most common slicer settings, including how to change printers, change materials, and access the quality settings menu. Then we learned about some of the most crucial settings to consider within the *Recommended* quality settings menu, including layer height profiles, infill density, support structures, and bed adhesion. Last, we wrapped up this chapter with some helpful tips and tricks to set your projects up for success.

7.12   Chapter Problems

- What is G-code? How does it link 3D printing with older technology, such as CNC lathes and mills?
- How do you set up a new 3D printer in Ultimaker Cura? Do you need to set up specific settings like build volume, G-code language, and boot instructions?
- Once you've imported a 3D file, how do you move it around in Cura?
- How do you rotate the object? Scale it?

- What is the standard order of operations if you want to duplicate an object?
- If you have many different small parts, do you need to arrange them each individually? Why or why not?
- If I wanted to 3D print with PETG instead of PLA, what settings would I change in Cura?
- What are the four most important settings to consider when first starting to 3D print?
- What does the *Profiles* (*layer height*) setting do? What are the pros and cons of high or low-resolution projects?
- What is infill? How does it affect the time and structural integrity of your project?
- What are supports?
- When might you want to toggle on supports?
- What types of build plate adhesion should you consider if you're printing many small parts with minimal surface area touching the build plate?
- What should you always make a habit of doing in *Preview* view before bringing your G-code file to the printer?

References

1. Ultimaker. Ultimaker Cura software web page. https://ultimaker.com/software/ultimaker-cura.
2. Prusa Research. Prusa Slicer software web page. https://www.prusa3d.com/page/prusaslicer_424/.
3. OctoPrint. OctoPrint software web page. https://octoprint.org/.
4. Slic3r. Slic3r software web page. https://slic3r.org/.

Advanced Slicer Settings

8

8.1   Objectives

Objectives: Now that you've become more familiar with slicing software, it's time to cover more advanced applications and custom settings within Cura. This chapter will review the additional settings that you might consider using to improve your 3D projects as your transition from novice to experienced user, including:

- Exploring the 12 major categories of Ultimaker Cura's *Custom Settings* panel:
  - Quality and layer height
  - Wall thickness and what it means
  - Top and bottom thickness
  - Infill density and pattern
  - Material temperature and when to tweak them
  - Print and travel speed to slow down or speed up
  - Travel and retraction settings
  - Cooling and fans
  - Support settings, pattern, and density to avoid support scarring
  - Build plate adhesion and line count
  - Special modes and why you might use them
  - Experimental settings
- Additional advice and recommendations for troubleshooting some of the most common problems you're likely to bump into as you grow more comfortable with 3D printing slicers.

© The Author(s), under exclusive license to Springer Nature Switzerland AG 2022   79
T. Kerr, *3D Printing*, Synthesis Lectures on Digital Circuits & Systems,
https://doi.org/10.1007/978-3-031-19350-7_8

8.2   Overview

In this chapter, we'll continue to utilize Ultimaker Cura since it can be used with a wide range of different desktop 3D printers [1]. Still, an understanding of general settings covered in this chapter, such as temperature, speed, travel, bed adhesion, and support settings should apply to most popular slicer software available today.

Cura's *Recommended* settings might be perfectly fine for the majority of your projects. As you grow more comfortable using slicers, however, you will likely encounter challenges that require a bit more customization to improve the print quality or resolve a particularly troublesome print failure. For example, maybe your print project is warping up at the corners of the model and not sticking well to the build plate. Perhaps the extruder leaves stringy, spiderweb-like wisps of material when it travels from point to point. Maybe the underside of your print has lots of support scarring where the supports contact the model, and you want a smoother bottom surface. Maybe you simply want to speed up your project's print time or improve its print quality. Knowing the purpose of these settings and how to change them allows you to customize almost every aspect of the 3D printing process. With some practice and a general understanding of what each setting does, each of these problems can be solved with the help of the *Custom* settings panel in Cura.

Cura's *Custom* settings panel allows you to change over 500 individual settings across 12 sections: Quality, Walls, Top/Bottom, Infill, Material, Speed, Travel, Cooling, Support, Build Plate Adhesion, Special Modes, and Experimental. These settings can give you considerable control over the 3D printer and the quality of your project. Better still, you can choose to show only the settings you want to adjust. To access the *Custom* settings panel, select the *Custom* button at the bottom of the *Recommended* settings panel (Fig. 8.1).

Once you've switched over to the *Custom* settings panel, you can expand or collapse any of the 12 section headers to view specific settings. Many settings within sections are not visible by default, so we'll cover the basics of navigating this advanced menu to add or remove settings from your *Custom* settings panel to suit your project. Then, we'll review what settings are included in each section, and how you might utilize them.

8.3   Cura's Custom Settings Panel

Cura's *Custom* settings panel will differ depending on the type of 3D printer you initially load into the slicer. Some printers may have dual extruders, requiring additional settings for both extruders. Your 3D printer may have a smaller printable volume, or may lack a heated build plate. The motors that control your 3D printer's three axes of movement likely differ from other 3D printer brands, which ultimately influences the layer height, print and travel speed, and many additional settings. Thus, the type of printer you have

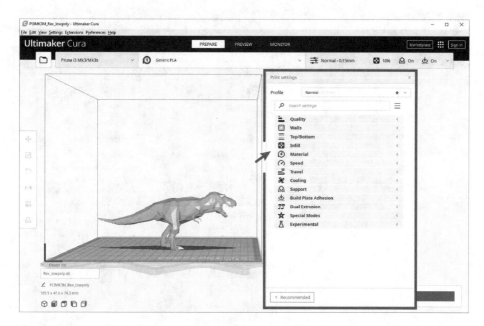

Fig. 8.1 Ultimaker Cura's *Custom* settings panel, highlighting the 12 expandable sections with customizable settings. Screenshot from Ultimaker Cura (https://ultimaker.com) software. One additional section (*Dual Extrusion*) does not apply to Prusa i3 MK3S+ 3D printers.

will strongly influence the settings available. The good news is that Cura factors in the parameters of your 3D printer and provides machine profiles that should automatically be imported when you load your 3D printer's make and model into the slicer. You can select one of these profiles under the *Profile* dropdown menu as an excellent starting point. Using these presets as a foundation, you can show or hide additional settings as you'd like. Whatever settings you add or remove from your Custom settings panel will still be there when you next open the Cura software.

Let's run through some of the most critical aspects of the *Custom* settings panel (Fig. 8.2).

1. The *profile dropdown* menu provides a list of preset profiles that you can choose from, typically ranging from very high resolution (0.06 mm) to very low or coarse resolution (0.4–0.6 mm). If you wish to create a custom profile of your own, you can save your current settings by selecting *Create profile from current settings/overrides*.
2. The *setting search bar* allows you to enter keywords to search for specific settings by name. Custom setting headers will collapse to show only the most relevant sections. Try searching for "support" to show all settings under the current setting profile related to that keyword.

Fig. 8.2 Navigating the
Custom settings panel.
Screenshot from Ultimaker
Cura (https://ultimaker.com)
software

3. The setting visibility menu allows you to swap between different setting profiles and the settings they make visible.
4. The *Basic* setting profile has far fewer settings visible compared to the *Expert* or *All* profiles. You can also select *Manage Setting Visibility* to manually toggle checkboxes to show or hide the most relevant settings for your project.
5. Similar to *Manage Setting Visibility* in the setting profile dropdown menu, you can select the filter icon when you hover over a setting category to show or hide the settings you need.
6. Each setting has setting parameters that you can change.
7. Setting parameters may have additional icons that indicate that they're linked to additional extruders (in the case of dual extruders) or that the setting is automatically calculated (f_x) by other settings. Some setting parameters may be grayed out and noninteractive, indicating that they're controlled by another setting. Hovering over any part of the setting parameter will display a tooltip with more information on what the setting does, what it may be affected by, and what it affects. You can also undo actions and reset the parameter to its default value.

8. Each of the 12 setting categories can be expanded or collapsed to tidy up your workspace.

Next, let's explore the specific settings that each section controls. We'll also review why you might want to change them to improve your project and potentially solve some common 3D printing problems. Remember, it's always good practice to change only one or two settings at a time. If your print project fails and you've changed too many settings, it can be difficult to determine which setting may have caused the failure. Instead, 3D printing should be an iterative process: change one or two settings to address the issue, and if the issue persists, change one or two more.

Ready to review the 12 *Custom* setting panel sections? First, let's navigate to the *setting visibility* menu and switch from *Basic* to *Advanced* mode so we can make a few more settings visible as we go. We'll cover many of the most relevant *Advanced Custom* setting panel settings throughout the rest of this chapter.

8.3.1   Quality

The quality of your 3D printed project is influenced by settings such as layer height and line width, and is easily one of the most important sections where you'll need to consider modifying settings. Recall from Chap. 7 that **layer height** refers to the height (thickness) of one layer of your print in millimeters (Fig. 8.3). Smaller (thinner) layers result in a higher-resolution print or smoother surfaces, but at the expense of time since they can take much longer to print. Larger (thicker) layers result in coarser prints, but often take only a fraction of the time to print. A good starting point is always 0.15 or 0.2 mm.

Initial layer height sets the height of the first layer of material that touches the build plate. Usually, this should be set to the same height or slightly thicker than the layer height value above. Why? More material commonly means a bit more surface area touching the build plate, which can help with bed adhesion.

Fig. 8.3 Left to right, a model at 0.06, 0.15, and 0.6 mm layer height. Note how the layer heights get thicker, and the model gets coarser. Screenshot from Ultimaker Cura (https://ultimaker.com) software

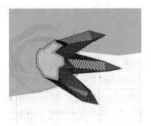  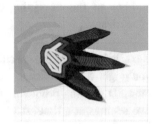

Fig. 8.4 Left to right, a model at 0.2, 0.4, and 0.8 mm line width. Like layer height, line width affects quality and speed of the project. See how the coarser model cannot resolve the sharp points on each toe of our *T. rex*? Screenshot from Ultimaker Cura (https://ultimaker.com) software

Line width dictates the width of a single line of filament as the nozzle moves along a path. Typically, the line width should be approximately the width of the nozzle, though you can reduce the line width value slightly to improve the quality of your print. Make sure you know what width your nozzle is (commonly 0.4 mm) and adjust this value accordingly. There are several sub-settings within line width that you can change that affect the width of outer and inner walls, infill, bed adhesion, supports, and even primer lines. Adjusting the line width will affect the quality of the model and speed of printing in a similar way to layer height, where thicker lines result in coarser details and faster prints, and thinner lines bring out finer detail and slower prints (Fig. 8.4).

8.3.2 Walls

Wall thickness can influence the strength and sturdiness of a 3D printed part, and potentially also how airtight and watertight the part is. Increasing **wall thickness** allows you to make a part sturdier without affecting the outer dimensions of the model, but will also increase the print time and often require more material. You can enter wall thickness parameters either as a number (which Cura rounds to a multiple of the line width) or as a specific number of walls (Fig. 8.5). Typically, multiplying the line width by two or three (in other words, a wall line count of two or three) is perfectly fine for most applications. The wall thickness and **wall line count** parameters are linked, so changing one will gray out and automatically change the other.

If tolerance is a critical consideration for your 3D printed project, you might also utilize horizontal expansion. FDM 3D printing often uses thermoplastics, which can be prone to deforming as they're heated up, extruded out a very hot nozzle, and then rapidly cooled. You can increase the horizontal expansion parameter to compensate for holes that are too large (increasing the X/Y size of the model) or decrease the horizontal expansion (decreasing the X/Y size of the model) to account for holes that may be too small.

Fig. 8.5 Left to right: one wall, two walls, and four walls, shown as outer walls (red), inner walls (green), and infill (yellow). Wall thickness influences the strength and sturdiness of a 3D print, but often requires more material and takes more time to print, which may affect cost of material. Screenshot from Ultimaker Cura (https://ultimaker.com) software

Finally, it's worth noting that FDM 3D printing isn't ideal for any part that must be leakproof. As we've learned, FDM 3D printers create 3D printed parts by laying thin layers of thermoplastic on top of other layers, which often means that most FDM 3D prints have very tiny micropores, gaps, or imperfections between layers that liquids and gases can pass through. Thickening the walls may stem or slow the leaks, but post-processing the parts with a sealant or chemical treatment is often required to make true airtight or watertight parts with FDM 3D printers. If you're interested in leakproof 3D prints, stereolithography (SLA) or material jetting (MJ) may be better suited for the task.

8.3.3   Top/Bottom

Modifying *Top/Bottom* settings in Cura allows you to improve the surface quality of both the bottom and top of the model, and also allows you to adjust the thickness of both the top and bottom of the 3D print. Similar to the *Walls* section of Cura, you can bump up the thickness of the top and bottom (roof and floor) to make the part sturdier and reduce gaps. Notably, just like with the *Walls* settings, increasing the number of top and bottom layers will also increase the print time and require more material.

Ironing is a relatively new addition that provides an exciting feature. With ironing toggled on, the hot nozzle will retrace its path over the top layer once printing is complete to 'iron out' and smooth that top layer.

8.3.4   Infill

Remember that infill is the latticework of crisscrossing support structure inside a 3D printed object that provides internal structural support. Accordingly, infill plays a significant role in determining the strength and rigidity of a 3D printed part and directly affects factors like print time and material used.

The **infill density** of the part determines how much plastic is used as internal support within the model (Fig. 8.6). A higher density means more plastic and often a smaller latticework, which makes the object stronger but takes more time to print and uses more material. Typically, a 20% density or lower is perfectly fine for most projects. For educational 3D prints, visual prints that will simply sit on display, or prints that won't have any type of force exerted on them, even 5% infill density should work well. If your 3D part will be undergoing significant stress and strain, however, you might consider bumping that infill density up to 20% or higher depending on your desired use.

Many slicers offer different **infill patterns** to select, which serve various purposes depending on your project's desired outcome (Fig. 8.7). Cura defaults to a grid pattern, which prints relatively quickly and provides sufficient strength. If your project must have specific mechanical properties to serve a specific purpose, you might want it to be strong in a particular direction or flexible in another. You can choose from a wider number of infill patterns, which Cura groups into four basic categories:

- *Strong 2D infills* (grid, triangles, tri-hexagon) are used for most standard prints. Grid is the default infill pattern, and suits the majority of projects.
- *Quick 2D infills* (lines, zig-zag, lightning) are used for models you want to print quickly at the expense of strength.
- 3D infills (Cubic, cubic subdivision, octet, quarter cubic, gyroid) make the object equally strong in all directions. Accordingly, they are best suited for functional objects that might undergo stress or strain.
- *3D concentric infills* (concentric, cross, cross 3D) are best used if you're using flexible materials.

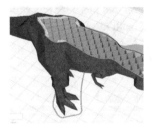

Fig. 8.6 Our *T. rex* model's infill. Left to right: 10, 20, and 40% infill density. Infill density is useful for strengthening 3D printed parts. Screenshot from Ultimaker Cura (https://ultimaker.com) software

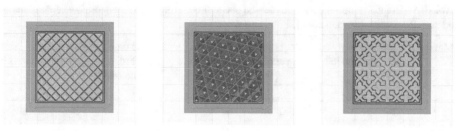

Fig. 8.7 Infill patterns. From left to right: grid (strong 3D infill), cubic (3D infill), and cross (3D concentric infill). Screenshot from Ultimaker Cura (https://ultimaker.com) software

Finally, *gradual infill steps* is an interesting and relatively new setting added to the infill menu that can be used to reduce print time significantly. **Gradual infill** works by splitting the infill density in half for every step indicated in the *gradual infill steps* parameter, with the lowest density infill nearest the bottom of the model. In contrast, areas closer to the model's top surface will approach the infill density parameter you've already set in Cura. The rationale here is that the roof of any model will require structural support, similar to the purpose of external supports that you may have toggled on. High-density supports are less crucial for the bottom of the model. Thus, in order for the top surface of the model to be laid down correctly, there must be sufficient infill to support the roof. By establishing a set number of gradual infill steps (say, two steps) as well as an infill density (say, 20%), the first section of the infill will be printed at 5% infill density, then 10% infill density, and closest to the top surface, 20% infill density. This comes at the expense of the model's overall strength, but can significantly speed up the print time of your project.

8.3.5   Material

Most types and brands of thermoplastic filament have specific material and temperature settings required for them to print properly. The **printing temperature** refers to the temperature of the nozzle while printing, whereas the **printing temperature initial layer** is the nozzle's temperature only while printing the first layer touching the build plate. Therefore, a slight increase to the *printing temperature initial layer* can help with bed adhesion.

Both *initial printing temperature* and *final printing temperature* are settings specific only to dual-extrusion machines and not currently relevant to Prusa 3D printers.

The **build plate temperature** controls the temperature of the heated build plate. Most 3D printers today have heated build plates, and those without heated build plates can be exceedingly difficult to dial in and get prints to adhere. **Build plate temperature initial layer** is the temperature of the build plate when printing the first layer. Just like

Table 8.1 Helpful material guide, condensed and sourced from https://help.prusa3d.com/materials © Prusa Research—prusa3d.com [2]

Material	Enclosure recommended?	Rec. nozzle temperature	Rec. bed temperature	Prints on a standard smooth build plate?	Glue stick recommended?
PLA	✗	210–215 °C	60 °C	✔	✗
PETG	✗	230–270 °C	90 °C	✔	✔
ASA	✔	260–265 °C	95–110 °C	✔	✔
ABS	✔	240–255 °C	110 °C	✔	✔
PC	✔	270–275 °C	115 °C	✔	✔
PVA	✗	195–215 °C	60 °C	✔	✗
TPU	✗	230–260 °C	50–85 °C	✔	✔
Nylon	✔	250–270 °C	75–90 °C	✗	✗

the *printing temperature initial layer*, increasing the build plate temperature slightly can help with bed adhesion and prevent warping. When fine-tuning temperature, it's good practice to move up or down in smaller increments.

Prusa provides a very helpful materials guide table, which we've distilled down to more common or popular filaments below (Table 8.1).

8.3.6 Speed

Printing speed refers to the speed at which the extruder travels as it lays down filament along walls, infills, and support structures and as it meanders from place to place. Speed is an essential factor to consider manually adjusting, as it can affect the total print time of a project and the quality of the print. Typically, Cura defaults to 60 mm/s, but the optimal speed settings you might choose are often influenced by the geometry and purpose of your project, in addition to the filament you're using, the 3D printer you've selected, and your desired layer height. Similar to temperature settings, it's a good idea to iterate and experiment with print speed settings in small, 5 mm/s increments.

In a world of rapid prototyping, it's reasonable to assume that most folks would prefer to print their objects quickly, but there are some important factors to consider. For example, increasing the print speed will result in a shorter print time but may affect quality and detail if the printer is moving too fast. Likewise, you may need to increase the temperature to ensure that the filament is melting properly, and is in sync with the flow rate of filament and movement of the extruder. If not, you might run into under-extrusion.

The **print speed** of a model refers to the speed (in mm/s) that the printer is moving while extruding filament. Under the *Advanced* setting profile, you can customize different print speeds for different parts of the model, such as infill, outer and inner walls, top/bottom surfaces, supports, bed adhesion, and initial layers. Slower speeds along outer walls and top and bottom surfaces ensure a better surface finish and that any gaps along the surface are closed. Faster infill speeds, particularly for models that don't require any type of strong internal support, can improve print time since it doesn't matter as much if the infill looks messy.

Travel speed refers to the speed of the extruder when moving around and not extruding material. While your nozzle sits at over 200 °C, there's always a chance that tiny wisps of filament ooze out from the nozzle. These are easily cleaned up, but may be a bit of a headache and worth avoiding if you can change settings to do so. Retraction and "Z-hop" settings that we'll cover under *Travel* usually help improve prints that show signs of this stringing and oozing, but adjusting travel speed can also play a part. If you increase the travel speed, you reduce the chance of the filament oozing out in the same place with every new layer, which can improve part quality. But with increased speed, there's an increased chance of the extruder bumping into parts of the print or negatively affecting quality by moving so quickly. So, just like many other settings, it's good to make small, iterative changes to travel speed. Many popular desktop FDM 3D printers are happiest traveling at 120–150 mm/s.

Finally, it should be no surprise that you can modify the **initial layer print speed** and the **skirt/brim speed**. If your print is having trouble sticking to the build plate and is peeling up or warping, or if your brim or skirt bed adhesion is not adhering well, you might consider slowing down the speed of the first few layers of the model and the surrounding bed adhesion. Sometimes slowing down the first few layers of a project can significantly improve the quality of the entire project by ensuring a solid, well-adhered foundation layer. Remember: 3D printing is a bit like building a house. Without a strong and supportive foundation, the rest of the house can't stand.

8.3.7   Travel

As your extruder zips back and forth, laying down filament, the heat from the nozzle and the constant flow of new filament through the hot end can lead to small stringy bits of filament oozing out. Often, these tiny stringy wisps of filament are part of the model, looking a bit like a spiderweb between points of the model (Fig. 8.8). Usually, this "**stringing**" is easy to remove, simply by pulling it off, scraping it off with a sharp blade, or even using a hairdryer. Still, there are settings within most slicers that allow you to mitigate this annoying stringing by effectively "retracting," or pulling back the filament as it travels and pushing it forward again during filament deposition.

Fig. 8.8 Examples of stringing on 3D printed projects. © Prusa Research—Print Quality Guide—prusa3d.com

Toggling on the *enable retraction* feature allows you to allow **retraction** as the extruder travels. This can prevent those thin threads of thermoplastic that may clump up between points of your print, making the 3D print a bit cleaner and requiring less post-processing. If you're using flexible TPU filament, however, be careful with retraction settings, as moving the filament back and forth through the cold end relatively rapidly may cause filament grinding or filament jams. A quick internet search of retraction settings for your specific flexible TPU filament should yield a number of recommended settings to help avoid this.

Within Cura, you can control the **retraction distance**, which is the distance in millimeters that the filament will be pulled back from the nozzle. Shorter retractions mean a higher chance of oozing out of the nozzle, but ensure that the material is less prone to jamming or grinding, and also shorten the print time of the model. Conversely, longer retraction distances may reduce oozing more effectively, but will add time to the printing process and present a higher risk for filament grinding or jamming.

Likewise, **retraction speed** controls the speed at which filament is pulled back and pushed forward by the hot end. Similar to retraction distance, lower speeds are safer against grinding and jamming, but increase the chance of oozing, whereas faster speeds open up more room for error but lower the risk of machine malfunction.

Suppose your 3D print is complex, and you are worried about the extruder bumping into and potentially knocking over parts of the model. In that case, you can consider toggling on features such as *combing mode*, *avoid printed parts when traveling*, and *avoid supports when traveling*. These all serve a similar purpose: ensuring that the extruder steers clear of certain parts of the model, or travels only around the perimeter of the model. However, because the extruder won't move in an optimized path, these settings may add additional time to your project.

Finally, you might wish to explore *z-hop when retracted*, which (depending on the z-axis on your 3D printer) allows the build plate or the hot end to move up or down when the extruder travels. This creates a bit more breathing room between nozzle and print, reducing the chance of the nozzle causing stringing or oozing.

8.3.8 Cooling

Cura's *Cooling* section allows you to manually control the hot end fans throughout printing, which can be hugely helpful if you're having challenges getting first layers to adhere to the build plate, or conversely if you're encountering issues with your part shrinking. Toggling on **enable print cooling** allows you to speed up or slow down the extruder and simultaneously ramp up or down the fan speed. For example, you might want to increase the fans to cool the print and reduce stringing, or you might want to lower the fan speed to improve bed adhesion. Most slicers allow you to manually adjust both settings.

Fan speed settings allow you to change the speed at which the cooling fans spin, starting at a rate of 100% speed. Fans spinning at a higher rate usually mean better cooling, less oozing, and a bit better part quality but can cause parts to shrink a bit. Fans spinning at a slower rate allow layers—either the initial layer against the build plate or two adjacent layers—to better stick together.

Several settings, such as **initial fan speed** and **regular fan speed at height** (or **regular fan speed at layer**), allow you to set a faster or slower fan rate for the first few layers. Accordingly, this fan speed will gradually increase as your part approaches the specified height (in mm) or numerical layer. *Regular fan speed at height* and *regular fan speed at layer* parameters are linked, so changing one will gray out and automatically change the other. Sometimes, setting the initial fan speed to zero allows the print to more effectively adhere to the build plate.

Finally, with *minimum layer time*, you can control how much time (in seconds) the extruder spends on a particular layer. This can be particularly useful if you want to allow one layer to cool before placing down another. If your part is small, the print spends less time on each layer, which ultimately means less time to cool between layers. Less time to cool often results in a poorer part quality as well as oozing and stringing. *Minimum speed* acts as a gatekeeper for *minimum layer time*, as you don't want the printer to slow down too much or risk poor part quality. Accordingly, *minimum speed* can override *minimum layer time* settings that would force the printer to slow down below an acceptable level. Toggling on *lift head* also addresses this problem by moving the hot end away from the layer until the minimum layer time has passed to allow the parts to cool sufficiently.

8.3.9 Support

There's a very good chance that a model you want to 3D print has overhanging parts. Recall from Chap. 7 that supports are disposable structures that break off or dissolve away from your model, and help to prop up any parts of the 3D model that lean or project out beyond a 45° to 65° angle during the printing process. The Y-H-T rule might help you to determine if supports are necessary: a Y-shaped object doesn't necessarily need supports, but H- and T-shaped objects usually require supports to properly print. Toggling on **generate supports** will open up a menu with several useful settings to change.

Support structure allows you to select from one of two primary types of supports: normal supports and tree supports. Normal supports are on by default and will produce a simple latticework of 'support beams' under your model's overhanging features. On the other hand, tree supports will wind and snake their way around like branches on a tree. You might consider these for models where you want to minimize support scarring, use less material (and thus take less time), or if your model has very complex geometry that normal supports might struggle to scaffold correctly.

You might want to consider whether you only need supports touching the build plate, or supports everywhere throughout the model where there are overhangs through the *support placement* parameter. If you wanted a bit more control to place supports only in areas touching the build plate, this is a great setting to adjust to minimize issues like support scarring, but increases the risk of printing unsupported structures.

As you grow more comfortable with 3D printing and as you start to better conceptualize how your object might 3D print, you may be able to push the boundaries of the **support overhang angle** (Fig. 8.9). This parameter allows you to adjust the minimum angle for which supports are added, where 0° will print supports on every overhang and 90° will not print any supports. The *support overhang angle* also affects how much support material is added: if you set the angle to 10°, much more material will be used compared to 65°or 85°. The more you practice 3D printing, the more your confidence will grow. You'll be able to recognize areas where Cura thinks you should have supports, but where you know you can adjust other settings in order to print without supports, or with fewer support structures.

Like infill, **support pattern** allows you to select different patterns that all have different benefits, from the strength of supports to the ease at which support material breaks away.

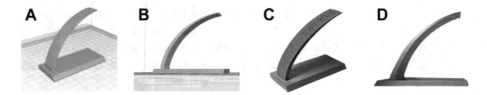

Fig. 8.9 Overhang angle, shown in Ultimaker Cura (https://ultimaker.com) software (**A, B**) and as a printed part (**C, D**). Notice the rough surface finish on a part printed without support (**D**) as the overhang angle increases. Notably, Cura suggests supports are needed around 50°, but the surface finish of the actual 3D printed part does not begin to diminish until around 65–70°

Fig. 8.10 3D printed Dungeons & Dragons miniatures can be difficult to print, as supports can be hard to remove from tiny features such as arms and legs. Can you spot the support scarring at the base of the red figurine's feet?

Similarly, you can control how dense the supports are through **support density**. Denser supports will be stronger but may be harder to remove. Moreover, denser supports printed between particularly complex parts of the model may be so hard to remove that they risk accidentally breaking off smaller, more fragile parts of the model (Fig. 8.10).

For more advanced settings, you might consider adjusting the *support horizontal expansion*, which allows you to offset the support structure along the X and Y planes. This effectively allows you to expand the support area to make a model sturdier. Support horizontal expansion can also contract the support area (by entering a negative value) to eliminate support structures that may only feature tiny areas of support, and that don't play a significant role in supporting the model. As with so many other settings, increasing or decreasing the support area significantly affects overall print time.

Since it's not mission-critical for support material to look beautiful and print at a high resolution, you can adjust the *support infill layer thickness* to speed up printing by printing two layers at a time. For example, if your layer height is set to 0.15 mm, you can consider printing supports at a thickness of 0.3 mm. Keeping this value as a multiple of the layer height is a good idea. Similar to your gradual infill setting parameter, you can also control the *gradual support infill steps* to speed up printing considerably.

Finally, you can toggle on *enable support interface* if you have significant issues with support scarring. Adding a support interface creates a breakaway roof or floor surface that—while more difficult to remove than traditional breakaway supports—creates a thin planar layer of material between the supports and the model. Think of it like a cushion between the smooth model and the rougher supports.

8.3.10 Build Plate Adhesion

In Chap. 7, we touched briefly on bed adhesion and discussed that toggling on build plate adhesion from the *Recommended* settings menu turns on a brim that surrounds your model like the brim of a hat. This, in turn, increases the surface area of the print touching the build plate and helps the part to better adhere to the build plate. In Cura's *Custom* settings, you can customize the type of build plate adhesion as well as the general width or line count of the build plate adhesion. There are three types of build plate adhesion that Cura offers:

A **skirt** is a line of filament that draw a perimeter around your object but does not touch the object itself. Skirts help to prime the nozzle so the machine can quickly start printing without errors. As the 3D printer's nozzle heats up, small amounts of filament might ooze out while the extruder idles in a corner and waits to reach the correct temperature. Your 3D printer is smart, but likely not quite smart enough to know that there may be a small air gap at the tip of the nozzle where that filament just dripped out. Just like a primer line or primer blob, skirts help to ensure that there's sufficient thermoplastic sitting at the very tip of the nozzle, ready to lay down the very first line of your model correctly. Without a skirt, brim, or primer line, your model might fail to print part of the initial layer by printing air where it should have extruded filament, which can cause a domino effect throughout the printing process. You can adjust the *skirt line count* to increase or decrease the number of lines drawn around your model. As an added bonus feature, skirts can also help you tell if your build plate around the model is level (Fig. 8.11). If the skirt line appears too squished or too round, it indicates that the build plate might not be level or that your nozzle height is off, and ultimately that your project may have difficulties sticking well.

Recall that **brims** add a single layer of material around the circumference of the model, and that they require some minor post-processing to peel off once the model is printed.

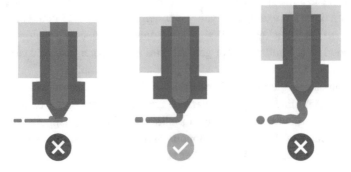

Fig. 8.11 How filament is laid down on the build plate is an indicator that your build plate or nozzle height may need adjustment

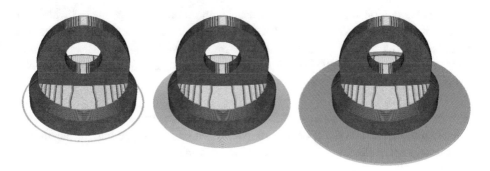

Fig. 8.12   Build plate adhesion types. Image from Ultimaker Support Guide (https://support.ultima ker.com/)

A brim's primary purpose is to increase surface area touching the build plate to improve adhesion and prevent issues like warping. As with *Wall* settings, under the *Advanced* setting profile, you can also control the numerical *brim width* (in mm) or *brim line count* to adjust the brim diameter. The brim width and brim line count parameters are linked, so changing one will gray out and automatically change the other. Models with tiny parts that all need to print together can benefit from a brim to help link them all together, but make post-processing and brim removal a bit tedious.

Finally, **rafts** print a thick horizontal platform of material between the model and the build plate. Rafts are more common for non-heated build plates, and since most modern 3D printers feature heated glass, metal, smooth, or textured build plates, rafts are not frequently used. Often, they can negatively affect the bottom floor surface finish of the model and make it appear very rough, almost appearing like support scarring. In specific use cases, you might consider adding a raft if your project has very few features touching the build plate and thus minimal area for brims or skirts to surround. Additionally, if your thermoplastic is prone to shrinking, such as ABS, you might consider using a raft or a brim (Fig. 8.12).

8.3.11   Special Modes

Modifying *Special Mode* settings in Cura allows you to control for some added customization in niche 3D printing use-cases. For example, *print sequence* determines whether or not the printer prints all models at once according to layer height, or each individual model on the build plate one at a time, provided there's enough room between models for the print head to move around and extrude filament safely. This might be used if you need to print one part faster than another, if you want to avoid things like stringing, or if you need to confirm that a part will print correctly before printing a multitude of them.

Surface mode is a bit more complex, but deals with non-manifold geometry, which you can essentially think of as geometry that can't exist in the real world. An easy example might be a 3D file you've downloaded online from a video game that has walls with no thickness. If you wanted to 3D print these, you could consider one of the three options in surface mode:

- *Normal* will print the model normally, but thinner geometry with no dimension will likely not print at all. You'll be able to quickly tell this when you hit *Preview* in your slicer and notice that some of the features are missing.
- *Surface* will turn every X and Y line into lines that are one nozzle-width thick, but will not print roofs or floors.
- *Both* will print the normal geometry, but will also print X and Y lines as one nozzle-width thick.

If you've experimented a bit with slicers while reading this book, you might have already heard of *spiralize outer contour*, which is often also called *vase mode*. This setting prints your model's outer wall very quickly as one single line, somewhat like making a soft serve ice cream cone. It's a particularly speedy way to 3D print with minimal material, but certainly doesn't produce strong models and thus is mainly used for visual, aesthetic 3D printing projects.

Finally, you can select the *Expert* setting profile to explore parameters such as *mold*, which creates a negative imprint of your model with a surrounding wall for you to use as a mold for other materials.

8.3.12 Experimental

Last up are the *Experimental* settings, a growing repository of creative, experimental settings developed by Ultimaker Cura and their partners and released to the public to beta test and provide feedback. There are a number of popular settings in Cura that initially started out as experimental settings. It's worth noting that since these experimental settings are frequently updated and published to Cura or removed, there may be some settings in the *Experimental* settings panel that are no longer there at the time of this book's publication.

Make overhang printable will convert all surfaces and geometry of your model that exceed the *maximum model angle*. This setting is hidden under the *Expert* setting profile to the value specified by *maximum model angle*. So if your maximum model angle is 65°, for example, a 77° angle slope and an 85° slope will be converted to 65° angles. This is an interesting setting to explore, particularly if you want to minimize supports, but will directly change the appearance of your 3D model.

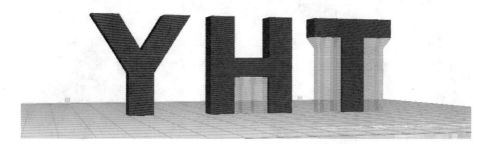

Fig. 8.13 Conical supports on T and H-shaped objects. Image from Ultimaker Cura (https://support.ultimaker.com/) software

If you have supports toggled on, *enable conical support* will appear in your *Experimental* settings panel (Fig. 8.13). This allows you to 3D print support structures that appear wider at the top and smaller at the bottom, like a cone or funnel. This can significantly reduce the support material needed, which can speed up 3D printing considerably.

Finally, **adaptive layers** can be a powerful setting that crunches some quick math to adjust the layer height of your model depending on its shape in real-time, as the model prints. With *adaptive layers* toggled on, Cura will analyze your model's geometry and optimize the layer height for each model section in a way that doesn't compromise part quality. As a result, curves and fine details will be printed at a finer layer height resolution than vertical, flat areas. In such a way, Cura can use varying layer heights to speed up print time and improve print quality. This feature is still in development, but increasingly popular across many different slicer software.

There is an ever-growing list of additional *Experimental* setting panel features to choose from if you select the *Expert* or *All* settings profiles that we won't detail here. Remember that hovering over any setting will provide a tooltip that provides excellent context on what a setting parameter does. As you grow more comfortable and confident with your 3D printing abilities, we encourage you to explore the large array of experimental settings yourself!

8.4   Troubleshooting Common Issues

Ultimately, it's important to remember that every 3D printer is different. Even if you purchase two identical 3D printer kits from the same company, you might adjust belts and tighten screws differently on one versus the other when you assemble them. All these factors can influence your final 3D printed project. While a bit of a spooky name, a good assessment of a 3D printer's health and capabilities are free, downloadable 3D

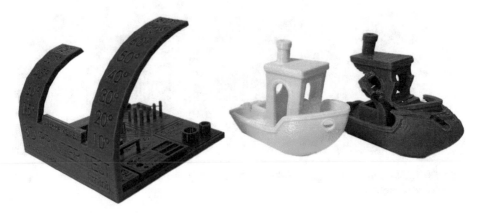

Fig. 8.14 Torture tests and calibration prints. Note the benchmark test (green) that failed spectacularly. Can you use the troubleshooting guide below to identify what went wrong?

printed parts called "torture tests" (Fig. 8.14) [3]. These torture test models are specifically designed to test the boundaries of your 3D printer, and should be printed according to the designer's specifications (usually things like "no supports" and "100% infill"). They can significantly help you calibrate and dial in your 3D printer according to how well it can print overhangs, dimensionally accurate parts, circular holes, small features, and how well it can span bridges, prevent stringing, handle shifts in layer quality, and so on. "Benchy," a cute tugboat 3D model designed by CreativeTools [4], is a popular benchmark torture test, though more comprehensive calibration tests are available.

Now that you have an in-depth understanding of slicer settings and what they do, it's time to provide a quick list of common problems and their solutions [5−8]. If you bump into any troubleshooting challenges while 3D printing, you can use the toolkit developed throughout Chaps. 7 and 8 to find creative solutions.

8.4.1   Print Not Sticking to the Build Plate

The first few layers of your print are easily the most important. If your print isn't sticking well to the build plate (Fig. 8.15), it could be a result of a few different issues, which means that there are a few solutions you might try:

Fig. 8.15 This print is not adhering properly to the build plate. Image from Ultimaker Support Guide (https://support. ultimaker.com/)

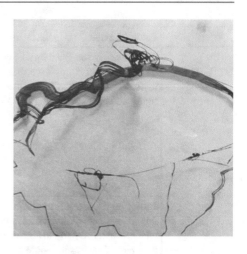

	Potential issue	Suggestions
See Fig. 8.15	**The build plate isn't level.** If the print bed isn't level, one area of your bed may be too far away from the nozzle, and another area may be too close	• Refer to your printer's manual to level the bed. Some printers have auto-leveling systems, and others require that you do it by hand
	The build plate is problematic. These days, most build plates are heated, and are typically made of smooth metal, glass, or unique textured material	• Make sure your build plate is free of dust and shows no signs of excess fingerprints or glue • If your build plate is not heated, consider laying down something like gaffer's tape or blue painter's tape to create more of a textured surface to help prints adhere • If your heated build plate is made of glass or metal, clean off the surface to get rid of any fingerprint oil, and apply a very thin layer of glue from a water-soluble glue stick (such as Elmer's washable school glue sticks)
	The initial layer is printing too fast. Each print layer must have time to cool and adequately bond with other surfaces	• Slow the *initial layer speed* down to allow more time to print each layer. For Prusas, 10 mm/s is relatively slow but can help parts better adhere
	The project lacks a brim or a raft. Brims and rafts can help increase surface area, and improve build plate adhesion	• By increasing the surface area, particularly for small parts, you increase the probability that the print will be able to find purchase and stick to the build plate

Potential issue	Suggestions
The nozzle is too far away. If your nozzle is too high up, it won't be able to correctly lay down filament, and the filament won't stick to the build plate	• Most printers allow you to adjust the height of the nozzle on the fly, even while printing. You can typically find this adjustment in your printers' settings • On a Prusa, this is found in the printer menu under *Calibration > Fist Layer Calibration.* You can also select *Live Adjust Z* during the printing process to make minor adjustments
Initial layer fans are too high. You don't want to cool the print *too* quickly, which can cause parts to shrink before they have time to settle onto the build plate	• Adjust the *initial fan speed* down, or disable it entirely for the first few layers • Set the *regular fan speed at layer* to ramp back up to normal fan speeds (typically 100%) around layer 5

8.4.2 Print is Warping or Peeling Off the Build Plate

Parts that warp or peel up at the corners (Fig. 8.16) are relatively common when 3D printing larger objects, or objects that take up more area across the build plate. To prevent **warping** or **peeling** and to ensure that your project adheres well to the build plate without shrinking, try these tips:

Fig. 8.16 A print warping off the bed. © Prusa Research—Print Quality Guide—prusa3d.com

	Potential issue	Suggestions
See Fig. 8.16	**The project lacks a brim or a raft.** Brims and rafts can help increase surface area, and prevent warping	• By increasing the surface area, you decrease the likelihood that the actual part itself will warp. Instead, parts of the brim or raft may warp slightly, which is less likely to affect your final project
	There is no adhesive on the build plate. Tape or glue can help increase the surface texture of the build plate, which can mitigate warping	• If your build plate is not heated, consider laying down something like gaffer's tape or blue painter's tape to create more of a textured surface to help prints adhere • If your heated build plate is made of glass or metal, clean off the surface to get rid of any fingerprint oil, and apply a very thin layer of glue from a water-soluble glue stick (such as Elmer's washable school glue sticks)
	The heated build plate is turned off. For most projects, ensuring that the initial layers of the print are warm enough to settle into the build plate is key	• For machines with heated build plates, ensure the build plate is on and heating properly. Recommended temperature settings are listed earlier in the *Materials* section of Chap. 8. Alternatively, search online for recommended settings for the specific filament brand you've purchased
	The fans are too high. Similar to prints that fail to stick to a build plate, prints can warp up if they cool too quickly	• Adjust the *initial fan speed* down or disable it entirely for the first few layers to allow the print to remain warm (and prevent shrinking) for longer • Set the *regular fan speed at layer* to ramp back up to normal fan speeds (typically 100%) around layer 5 • For materials such as ABS that are particularly prone to warping, consider turning off fans for the entirety of the print

Potential issue	Suggestions
There is too much temperature fluctuation around the 3D printer. Creating an environment that can maintain a controlled temperature is crucial for 3D printing	• 3D printers situated in breezy areas or areas with high airflow will be particularly prone to warping • While heated build plates help to regulate the temperature of the first few layers, you might need to better control temperature of the entire print—particularly if the print is large or takes a long time. When this plastic cools and shrinks, it may warp • You can buy or build a permanent heated enclosure out of acrylic, wood, or plexiglass. In a pinch, consider making one out of cardboard, foil, or other materials. If it keeps the heat in and the breezes out, it should work

8.4.3   Print is Stringing or Oozing

Recall that as your extruder travels across the build plate, the hot nozzle can leak tiny stringy bits of filament (Fig. 8.17). This spiderweb-like **stringing** is easy to remove but may still be a headache for those who want high-quality prints without much post-processing. Several slicer settings allow you to reduce the amount of stringing in a project:

Fig. 8.17 Stringing. Image from Ultimaker Support Guide (https://support.ultimaker.com/)

	Potential issue	Suggestions
See Fig. 8.17	**Your printing temperature is too high.** Higher temperatures can result in a tiny amount of filament oozing out of the nozzle	• Try adjusting the printing temperature down in 5°or 10° increments. Too hot can lead to increased oozing and stringing, but too cool can lead to clogs or jams. It's best to adjust the temperature in small, iterative steps
	Your travel speed is too slow. Quicker travel between parts means fewer opportunities for stringing	• You can increase *travel speed* under the *Speed* setting panel. Remember that with increased speed, there's an increased chance of the extruder bumping into parts of the print or negatively affecting quality because of this faster movement
	Your retraction settings need adjustment. Controlling the speed, distance, and Z-hop of filament can help mitigate stringing and oozing	• Increasing *retraction distance* under the *Travel* settings panel may reduce oozing and stringing, but increases print time and risk of filament jams • Increasing *retraction speed* can help prevent stringing, but may increase the likelihood of filament breaking or jamming • Toggling on *z-hop when retracted* allows the build plate or the hot end to move up or down when the extruder travels, which increases the distance between the nozzle and the print and reduces the chance of stringing or oozing • You should be able to look up recommended retraction settings for your printer and your material. Direct drive and Bowden extruders require different retraction settings, as do different materials such as PETG or TPU
	You may need to explore Special Mode *Print Sequence* **settings.** Printing each individual part on a build plate separate reduces travel between prints	• Remember that *Print Sequence* allows you to print each part on a build plate individually. You can reduce opportunities for stringing if the parts are created individually

8.4.4 Print Has Shifted During Printing

Layer shifts while 3D printing (Fig. 8.18) can be hard to diagnose and troubleshoot, as this misalignment can be caused by slicer settings or mechanical issues with your 3D printer itself. Since few (if any) commercial desktop 3D printers have sensors that may tell them that the extruder is off course, layer shifts won't be detected, and the printer will continue printing as if nothing has happened. Bumping the printer or the table that the printer sits on can sometimes cause this. While slicer settings are simple enough to adjust, mechanical issues require more time and energy to resolve. If you notice that your 3D printed project has layers that have shifted or moved significantly, try these steps:

	Potential issue	Suggestions
See Fig. 8.18	**You need to slow down the travel and print speeds.** When you print at higher speeds, your hot end may try to move faster than the motors can handle	• If your print is still in-progress, observe your machine. If you hear a clicking sound from your motors, this could indicate that they're moving too quickly • You might try restarting your print one more time. If the error occurs again, reduce your project's travel speed and print speed by 10 mm/s increments

Fig. 8.18 An example of layer shifting, where the print has jumped forward. © Prusa Research—Print Quality Guide—prusa3d.com

Potential issue	Suggestions
Your printer's belts need to be inspected and serviced. Over time, 3D printer belts may loosen and slip. Alternatively, belts may be too tight to allow free movement of critical bearings and motors	• Most 3D printers use belts made of reinforced rubber to control the position of either the hot end, build plate, or both. With extended use, these belts may become a bit loose. When properly tightened, most belts should sound a bit like an out-of-tune guitar • Follow your machine manual to learn how to tighten these belts to prevent them from slipping • Alternatively, if belts are too tight, loosen them to an acceptable level to reduce friction and allow bearings and pulleys to move more freely
Your linear rods need to be inspected and serviced. Linear rods or rails help guide a 3D printer's moving parts along an axis, and must be well lubricated	• Just like belts, linear rods or rails are critical to the function of a 3D printer. The more you use your printer, the more likely it is that gunk may build up along them • If these aren't well lubricated, friction can cause the extruder to get stuck or skip. Wipe linear rods down and reapply machine oil to address this
Your lead axis screws need to be inspected and serviced. Threaded rods or lead screws connected to stepper motors often drive specific axes of movement and must be greased	• Lead axis screws or threaded rods driven by motors are the primary means by which an axis moves. On a Prusa print, the Z-axis is driven by a left and right threaded rod • Just like linear rods or rails, make sure that these are greased well to allow proper motion. Instead of machine oil, your printer likely came with a small container of grease that you can apply to the length of the lead axis screw. Refer to your machine manual to do so

8.4.5   Print is Under-Extruding or Not Extruding Enough Material

Your 3D printer should extrude thermoplastic filament at a steady rate as it travels around the build plate. If your printer is not extruding enough filament (Fig. 8.19), you may notice gaps in the walls of your model, or between layers. This **under-extrusion** can result in a very poor print quality, and significantly weaker prints, and will likely mean that your print has failed. Try some of these tips to fix it:

	Potential issue	Suggestions
See Fig. 8.19	**Nozzle is clogged.** If your nozzle is partially clogged, insufficient filament will be extruded	• If there is debris in your nozzle, it may partially clog your nozzle. A good indicator of debris is filament extruding at a sharp angle during the filament loading step. As you load new filament, old filament sitting in the nozzle reservoir should eject out of the nozzle quite quickly. Watch this changeover to assess if the nozzle is partially clogged • Follow your machine manual for clogged nozzles. This can include hot ("atomic") or cold pulls, specialty cleaning filament, or even acupuncture tools to dislodge clogged material • Use brass brushes to gently clean the outside of the nozzle off as it sits at temperature, being *very* careful not to touch or damage the thermistors

Fig. 8.19 This project displays increasing under-extrusion as the print progresses

Potential issue	Suggestions
Filament is tangled or stripped. If your filament spool is tangled or stripped and cannot grip and drive more filament to the hot end, the printer will continue printing but may under-extrude or ultimately stop extruding	• Check your spool for tangles. Typically, spools that come directly from the manufacturer shouldn't be tangled • Always hold filament tightly and thread it through any spool holders to prevent it from uncoiling after use • Filament that's getting harder to unspool due to a tangle down the line will appear like increasingly under-extruded layers
Filament diameter is incorrect. If you use the wrong filament, or filament from an unreliable vendor, it may not be the proper uniform diameter through the spool	• The filament diameter should match the printer's specifications. Prusa printers use 1.75 mm thick filament • If you suspect that your filament may be poor quality and inconsistent throughout the length of the filament, you might want to check. Use calipers at different areas along the length of the filament to determine how much it may vary. Filament should only vary ±0.05 mm throughout (for Prusas, 1.70–1.80 mm) • Though not common, you can add new filament settings or adjust existing presets by navigating along the top menu to *Settings > Extruder > Material > Manage Materials*. You can also adjust the flow rate (below)
Flow rate is insufficient. The flow rate of your material is typically set to 100%	• Your *Material > Flow* settings determine the speed at which the cold end motor pushes filament through your 3D printer. These are generally set to 100% but may differ depending on your filament type • You can increase these in small increments of 5–10% to see if it resolves the issue

8.4.6   Print Has Support Scarring

It's not uncommon for projects that produce support material out of the same material as the model to exhibit **support scarring** at the spots where the support has been broken away (Fig. 8.20). This support scarring can be a bit of a headache, and can negatively affect the appearance of your final 3D printed project. To reduce support scarring, here are a few tips:

Fig. 8.20 This fossil crocodile
skull was printed upside down
with incorrect support setting,
which caused significant
scarring

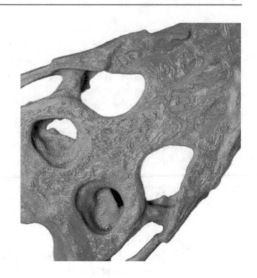

	Potential issue	Suggestions
See Fig. 8.20	**You aren't using soluble supports.** For 3D printers with more than one extruder, water-soluble supports easily address issues of support scarring	• Water-soluble supports dissolve in water over time and thus can be printed directly against the model. When they have dissolved away, they rarely leave any support scarring
	You need to adjust your support Z distance. Too close or too far away can both lead to support scarring	• Under Cura's *Expert* setting profile, the *Support Z Distance* refers to the distance from the top or bottom of the support structure to the print. The gap creates a bit of clearance to break off supports from the model • This can be tricky to dial in. Too close to the model, and it will be difficult to remove, potentially leaving some material behind that can scar the print. Too far away, and you have drooping overhangs. Start with a Z distance equal to the layer height and change by small increments. The Z distance will be rounded to a multiple of your specified layer height
	Your layer height is too high. Reducing your layer height means your printer can make smaller steps and fill in missing gaps	• If you lower the layer height of your project, your printer should be able to print finer details and thus print a bit closer to the model

Potential issue	Suggestions
Your support density is too low. Too much distance between adjacent support structures means more areas for overhangs to droop	• By increasing the *Support Density* percentage, you can usually get better overhangs. The tradeoff here is that the supports may be harder to remove
You need a support interface. When all else fails, adding a support interface creates a horizontal breakaway layer between the supports and your model	• Toggle on *Enable Support Interface* to add a cushion between the supports and your model. This breakaway cushion will be more challenging to remove but creates a thin planar layer of material between the supports and the model to minimize scarring. Worth noting: it will take more time for the part to print

8.4.7   Print Has Undesired Wavy Lines or Ripples on Surfaces

Small consistent ripples may be visible on your 3D print, which are good indicators that your machine may need tuning, your settings may need tweaking, or you might need to physically move your printer. This "**ringing**" effect appears like a wavy pattern across your printer (Fig. 8.21), most commonly due to the extruder's inertia as it travels around. To minimize these wavy patterns, try a few of the following:

Fig. 8.21 This project shows some evidence of ringing

	Potential issue	Suggestions
See Fig. 8.21	**Your printer has loose belts.** If the belts are loose, it can cause imprecise movements	• Tighten the 3D printer's belts until they sound like an out-of-tune guitar. Refer to your machine manual to do so
	Your printer is moving or accelerating too fast. The faster a printer moves, the more likely it is to vibrate	• When 3D printers make abrupt stops or direction changes, it can cause vibrations across the frame that create "ringing." Try slowing down the print and travel speed by small increments • Experiment with acceleration settings
	Your printer is not sitting on a stable surface. If your printer sits on a wobbly surface, the extruder will also wobble	• Ensure that your printer is sitting on a firm, stable surface or has something underneath (such as a cement block) to dampen any vibrations. If the 3D printer shakes the table, that wobble will carry over to the extruder

8.4.8   3D Printer is Clogged

A clogged nozzle is almost inevitable in 3D printing. The more you use your machine, the more likely small bits of thermoplastic get left behind inside the extruder (Fig. 8.22). Usually, **clogs** and **jams** are a result of dirt, debris, or heat creep. Read on to learn about a few solutions:

Fig. 8.22 This project highlights clear signs of a clogged nozzle. © Prusa Research—Print Quality Guide—prusa3d.com

	Potential issue	Suggestions
See Fig. 8.22	**There is debris inside the nozzle.** If the nozzle is clogged or partially clogged, debris may block the free flow of new thermoplastic	• As a first approach, you should consider performing a series of hot ("atomic") or cold pulls. This process might differ across 3D printers, so refer to your machine manual or reference online video guides specific to your make and model • Try simply manually feeding through more filament. Sometimes the only thing stuck filament needs is a little push. Your 3D printer's settings usually has a means to heat the nozzle and manually run filament through at a set rate • Alternatively, you can use special cleaning filament that is abrasive (but not abrasive enough to degrade the brass nozzle) to scrape away material • Finally, you can use acupuncture tools to poke up into the brass nozzle to clear debris. Unload your filament, heat the nozzle to temperature, and use the small acupuncture pins (or any very thin pin) to wiggle around in the nozzle
	Your 3D printer was sitting at temperature too long without moving filament. Heat creep is when the heat sink can longer effectively dissipate heat effectively, and filament begins to heat up and expand higher up along the hot end	• It's always a good habit to only let your printer sit at temperature when feeding filament at a steady rate through the machine. When the printer sits idle at temperature for too long, the heat block and heat sink can no longer effectively dissipate the heat. This can lead to filament that is higher up in the extruder heating up and expanding, which can seriously clog your print head. When not in use, make sure your extruder is not at temperature
	The outside of the nozzle is too dirty. If filament can't extrude properly because of older or crusty filament on the outside of the nozzle, it might under-extrude or create clogs	• Use a brass brush to gently clean the outside of the nozzle off as it sits at temperature, being very careful not to touch or damage the thermistors. Turn off the heat when you're done

8.4.9   3D Printer Has Stopped Midway Through the Project

Your 3D printer might have special hardware to detect errors in the printing process. In these cases, the printer will likely stop until the issue is resolved (Fig. 8.23). If your printer stops mid-print, issues might include:

	Potential issue	Suggestions
See Fig. 8.23	**The printer is out of filament.** Some 3D printers may be able to detect when filament runs out	• Your printer may be able to detect if you're out of filament, and will prompt you to load more before continuing. Follow the directions to load filament and resume your project
	The printer is waiting for you to change material. Special Cura settings allow you to change material on a single extruder machine at specified times	• Some 3D printers allow you to pause a print to change filaments if you wish to change colors or material at a particular layer height • In Cura, you can pause at height. The printer will wait for you to change filament and select continue before proceeding
	The printer has lost power. In the rare event of a power outage, some printers can create a save state to allow you to resume printing when you return	• More and more modern 3D printers have features that allow the machine to fully recover and resume prints after power outages. On Prusa machines, this is the Power Panic setting • Simply hit continue to resume your project, but watch it for about ten minutes to ensure the print looks okay

Fig. 8.23 A printer stopping mid-print usually indicates power loss, material issues, a clogged nozzle, or an intentional pause based on user settings

Potential issue	Suggestions
The filament has been stripped by the hobbed bolt. If your filament is tangled or if your idler tension needs adjustment, filament may be totally stripped and unable to get purchase against the idler and hobbed bolt	• It's pretty common for filament to become stripped. As the hobbed bolt rotates against the filament, it should be gripping and moving that filament up toward the hot end. If the filament is tangled or pushed too hard against the hobbed bolt, the hobbed bolt will dig in and grind away at the filament until there's nothing left to grab • Follow the filament grinding section below to fix this
The nozzle is clogged. The printer may detect a clog, and stop the project until the clog is resolved	• You'll need to unclog the nozzle, which may require removing your print, manually moving the extruder, and following instructions to free the clog. In these cases, you will likely need to restart your project. Follow the clogged nozzle instructions above

8.4.10 3D Printer Filament is Grinding

Most 3D printers use some combination of idler bearings to provide tension and hobbed bolts to provide traction as filament is fed into the cold end of the 3D printer and carried on to the hot end. In some cases, if the filament is unable to move because of a jam or because of too much pressure sandwiching the filament against the spinning hobbed bolt or drive gear, the cold end will grind away the filament until there is no more filament left to provide purchase (Fig. 8.24). In the case of **filament grinding**, consider one of the following solutions:

Fig. 8.24 Filament grinding is a common problem with a few easy fixes

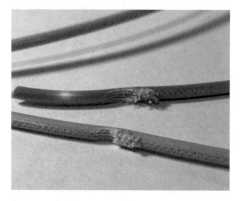

	Potential issue	Suggestions
See Fig. 8.24	**The idler tension is too high.** If the idler is pressing against the hobbed bolt too hard, it may grind filament down	• Too much tension against a hobbed bolt will cause filament to grind against its teeth. Most 3D printers have the ability to adjust this tension via a screw. Try reducing the tension by small increments and see if the problem is resolved • Make sure to also unload filament and blast compressed air toward the hobbed bolt teeth to ensure that the teeth are clean and can grip filament
	The hobbed bolt is dirty. The hobbed bolt has a series of small teeth to help it grip and pull filament. If these have plastic embedded in the teeth, filament will not gain any purchase	• Unload filament and blast compressed air toward the hobbed bolt teeth. Make sure to rotate the hobbed bolt to get all sides • Alternatively, you can use a brush to clean the hobbed bolt
	The nozzle is clogged. If filament can't fully move through the hot end, the filament will be stuck in place against a revolving hobbed bolt	• Follow the direction to clear the clogged nozzle above • Make sure to also unload filament and blast compressed air toward the hobbed bolt teeth to ensure that the teeth are clean and can grip filament
	Filament is tangled. If filament cannot be pulled into the machine, the filament currently in the cold end will grind down	• Carefully untangle the spool and then secure the end of the filament into the spool's spool holder • Make sure to also unload filament and blast compressed air toward the hobbed bolt teeth to ensure that the teeth are clean and can grip filament
	The temperature of the nozzle is too low. If the filament is not being sufficiently melted, it won't be able to extrude effectively	• Plastic will flow better at hotter temperatures. Try to ramp up the *Printing Temperature* of the material by 5 to 10°to ensure that the material is hot enough to melt and extrude out of the hot nozzle • Make sure to also unload filament and blast compressed air toward the hobbed bolt teeth to ensure that the teeth are clean and can grip filament

Potential issue	Suggestions
Retraction settings are too fast or too far. If your filament is being pulled back and forth too rapidly, it may grind filament	• If the printer moves material through the cold end too quickly, it can cause excess grinding. Try to drop the retraction speed by 50% and examine if the issue is resolved • Make sure to also unload filament and blast compressed air toward the hobbed bolt teeth to ensure that the teeth are clean and can grip filament
Print speed needs to be adjusted. If filament is moved through too quickly (or too slowly), it can cause excess wear and tear on the filament by the hobbed bolt	• Slowing down the print and travel speeds ensure that the motors aren't overworked. When they move at a slower rate, there's less risk of added strain to the cold end • Make sure to also unload filament and blast compressed air toward the hobbed bolt teeth to ensure that it is clean and can grip filament

8.4.11 Extruder is Moving Erratically

You'll usually hear mechanical failures before you see them, and they often require quick action to turn off the 3D printer and assess the damage. If the printer is unable to move, making loud mechanical noises as the motors turn in place, or is moving in an erratic, unpredictable manner, it can cause significant and sometimes permanent damage to your 3D printer's motors (Fig. 8.25). Make sure you're familiar with the following tips and tricks to prevent long-term damage to your machine:

Fig. 8.25 If your printer's
endstops break, the printer
won't know when the machine
should stop moving. © Prusa
Research—prusa3d.com

	Potential issue	Suggestions
See Fig. 8.25	**Slicer settings are incorrect.** If you select the wrong 3D printer, the build plate dimensions, G-code flavor, and printer settings will not match the current machine	• Most of the time, your printer won't allow you to print if you accidentally load settings for a different machine • In the rare case that the machine still operates with the incorrect G-code, you should immediately turn off the machine to prevent damage and upload the correct slicer settings
	The 3D printer's endstops are broken. Endstops are mechanical safeguards that tell the extruder or build plate that it has reached the maximum boundaries of its working area	• The endstops function like small bumper switches. When the printer runs into one, it triggers a mechanical switch to tell the machine it has reached the end of its motion • Endstop failure is usually pretty obvious, as the machine will make lots of unhappy mechanical whirring noises as it tries to move in the direction that the endstop failed to stop • Cycle on and off the machine. If the issue isn't fixed, you may need to replace your endstops

Potential issue	Suggestions
Printer is making a loud mechanical whirring noise. Immediately turn off your machine. Grinding motors are very bad for printer longevity	• Any crushing, rumbling, grinding, or abnormal noise coming from your motor should be treated as an immediate alarm bell. Immediately turn off your 3D printer • Grinding motors often indicate that an endstop is not working, that a rotating part is not operating correctly, or that a build plate sensor is not functioning properly and causing the nozzle to crash into the build plate • If you hear loud noises coming from your printer, immediately turn it off at the switch and assess what may be causing the issue
Firmware needs to be updated. It's a good habit to update your firmware every so often	• This is particularly true if you've recently purchased the printer, as firmware updates may have been published when your printer was already packaged and sitting in the warehouse • This is equally true if your printer is a bit older and hasn't been updated recently, as you will usually have to be diligent about updating it • Follow your machine's manual to flash firmware to your 3D printer

8.4.12 Nozzle is Scraping the Build Plate

Your build plate must always be level, free of debris and oil, and free of significant scratches or scrapes that prevent filament from being laid down in a predictable, even layer (Fig. 8.26). Nozzles that are too close to the build plate or crashing into the build plate because of a G-code or mechanical error can cause long-lasting damage, and may even require that you replace the build plate entirely. This may not be cost-prohibitive for some 3D printers with magnetic, glass, or swappable build plates, but it can be a pricey fix for other 3D printers with more permanently installed build plates. It's a good idea to adhere to the following in case of a nozzle impacting the build plate:

Fig. 8.26 Marks like these indicate that the nozzle may have accidentally scraped the build plate at one point

	Potential issue	Suggestions
See Fig. 8.26	**Your nozzle is too close to the build plate.** The z-height of the nozzle can usually be easily adjusted before and even during the printing process	• Adjusting the z-height or z-offset of the nozzle varies by printer. Refer to your machine manual to learn how to make adjustments • On a Prusa, you can find the proper settings to adjust the z-height of the nozzle in the printer menu under *Calibration > Fist Layer Calibration.* You can also select *Live Adjust Z* during the printing process to make minor adjustments if you are worried about your nozzle getting too close to the build plate
	Your build plate is not level. Any type of unlevel build plate will cause the nozzle to crash into the build plate	• Ensure your build plate has been leveled appropriately across the entire bed. This can be done by hand or by automatic leveling if your printer has specific settings to do so
	Your bed leveling sensor or endstop may be broken. A broken sensor will mean that the hot end will not properly detect the build plate	• You may need to replace the bed leveling sensor or endstop • On Prusa printers, the bed leveling sensor called the PINDA probe. The probe allows the extruder to detect how far the nozzle is from the build plate using induction. It senses when the extruder is close to a metal object (in Prusa's case, the metal build plate)

Potential issue	Suggestions
Your slicer's initial layer height settings may be incorrect. The settings entered into the slicer may not match the desired z-height settings determined by the stepper motors	• Initial layer height should usually be 50% to 80% of the layer height to ensure thermoplastic is effectively squishing into the build plate. Any less than this, and you might risk scraping the nozzle into the build plate • Check your slicer settings and adjust the initial layer height as necessary

8.5 Summary

In this chapter, we took a very deep dive into the advanced *Custom* settings that Ultimaker Cura offers. We first reviewed Cura's *Recommended* versus *Custom* settings, and detailed common features of the *Custom* settings menu. Then we explored each of the 12 primary sections of this advanced menu. While we didn't cover all 500 individual settings available, we reviewed the *Basic* and *Advanced* setting profile options to provide you with a sufficient understanding of how you might adjust settings to match the goals of your project. In the *Quality* settings section, we discussed ways in which you might improve the quality, surface finish, strength, and even bed adhesion of the part, as well as ways to speed up or slow down the printing process. In the *Walls* section, we reviewed ways to strengthen parts and even explored making them airtight and watertight. Under the *Top/Bottom* section, we learned a bit about how to improve the roof and floor surface finish. Within *Infill*, we learned about the pros and cons of different infill densities and patterns, and also learned ways to potentially speed up our project through gradual infill steps. In the *Material* section, we discussed the ways in which temperature plays a vital role in the success of a project, and reviewed a material settings guide for many popular thermoplastics. Under the *Speed* section, we reviewed how speed and acceleration of the extruder factor into print time and quality of the project, and talked about ways to improve initial layer adhesion and issues like warping. Within *Travel*, we learned about retraction and how to prevent stringing and oozing. In the *Cooling* section, we found out ways to regulate the fan speed and even the extruder's speed to allow more cooling between layers. Under the *Supports* section, we discovered ways to change support structure, density, overhang angle, and pattern. In the *Build Plate Adhesion* section, we touched on types of build plate adhesion and what each is well suited for. Finally, under both the *Special Modes* and *Experimental* sections, we explored some of the newer options available to improve or experiment with project settings. We then ended the chapter by discussing 12 of the most likely scenarios you might encounter that would cause your 3D printed project to fail, and how you might address and prevent them.

8.6   Chapter Problems

- How many custom setting sections does Cura provide? Approximately how many settings can you change?
- Why might you change the layer height of a project? What does changing the initial layer height in Cura do?
- Why might you increase the wall line count?
- What is a good infill density for a project that will undergo significant stress and strain? What's a good infill pattern for such a project?
- How does gradual infill play a role in the speed of a project? Does it affect strength?
- What recommended material settings do you need to print with TPU?
- Why might you slow down the initial layer speed of your project?
- What is retraction? How does it affect issues like stringing?
- Why might you set the initial fan speed in Cura to zero?
- When should you toggle on supports?
- What are the key differences and use cases for skirts, brims, and rafts?
- Can you identify any *Special Mode* or *Experimental* settings that might suit your specific project?

References

1. Ultimaker. Ultimaker Cura software web page. https://ultimaker.com/software/ultimaker-cura.
2. Prusa Material Table web page. https://help.prusa3d.com/materials.
3. Hullette, T. & O'Connell, J. 3D Printer Test Print: 15 Best 3D Printer Test Models. All3DP (2022).
4. Creative Tools. 3DBenchy homepage. https://www.3dbenchy.com/.
5. Jennings, A. 3D Printing Troubleshooting Common 3D Printing Problems. All3DP (2021).
6. Simplify3D. Print Quality Troubleshooting Guide web page. https://www.simplify3d.com/support/print-quality-troubleshooting/.
7. Prusa Research. Troubleshooting web page. https://help.prusa3d.com/category/troubleshooting_194.
8. Ultimaker. Ultimaker Support web page. https://support.ultimaker.com/.

Preparing to Print

9.1 Objectives

Objectives: We're in the home stretch now. This chapter will review the last few steps necessary to consider as you prepare your project for the 3D printer. We'll run through:

- What to double-check before printing
- Navigating the Prusa menu
- Best practices for loading filament
- Starting your 3D print
- Unloading and storing filament
- Best practices for most 3D printing projects

9.2 General Overview

After you've sliced your file, the *Slice* button is replaced by a *Save to Disk* or *Save to File* button. Use this button to save your project as G-code instructions to send to the printers, and also ensure that you've saved an editable project file (*File > Save Project*) in case you need to make changes. Once you have your file prepared in your slicer of choice, and once you've saved your project and exported your G-code in the proper format onto an SD card or USB to plug into the 3D printer, it's time to get the printer ready to go. The final section of this chapter will give you the tools necessary to print your project. You're in the home stretch now!

© The Author(s), under exclusive license to Springer Nature Switzerland AG 2022
T. Kerr, *3D Printing*, Synthesis Lectures on Digital Circuits & Systems,
https://doi.org/10.1007/978-3-031-19350-7_9

9.2.1   Double-Checking Your Slicer Settings

As you prepare your project for the 3D printers, it's a good idea to run through a quick mental checklist to ensure that your slicer settings work for the goals you wish to achieve:

- Does my project need supports? If so, have I toggled supports on?
- Does my project have small features touching the build plate? Is there minimal surface area touching the build plate? If so, have I toggled on "Bed Adhesion?"
- Have I sliced and viewed my project in *Preview* mode? Do I feel confident the first layers are oriented correctly and touching the build plate?
- Is the project set to the desired resolution and infill percentage?
- Is the project saved as a G-code file for the 3D printers in addition to a project file in case I need to make edits?

9.2.2   Navigating the Menu

Once you've saved your files, it's time to load filament, find your saved file, and start your project. To do so, simply plug in the SD card on the left of the LCD menu of the Prusa [1]. It may take a second to load the projects.

Controlling the LCD screen and navigating the menu is done by a single element: a rotational LCD knob that can be pressed to confirm the selection (Fig. 9.1). Press the LCD knob to enter the main menu.

When the printer is active, the main LCD panel (Fig. 9.1) provides the following information:

1. Nozzle temperature (actual/desired temperature)
2. Build plate temperature (actual/desired temperature)
3. Progress of printing in %—shown only during the printing
4. Status bar (Prusa i3 MK3S+ ready./Heating/file_name.gcode, etc.)
5. Z-axis position
6. Printing speed
7. Remaining time estimation

It's a good habit to always double-check your settings to ensure that the nozzle and build plate temperatures are accurate.

For now, you'll want to navigate to *Preheat* and select your desired material. It may take about ten minutes to heat up and should beep when ready. While the printer is heating up, you can pick out the material you wish to use.

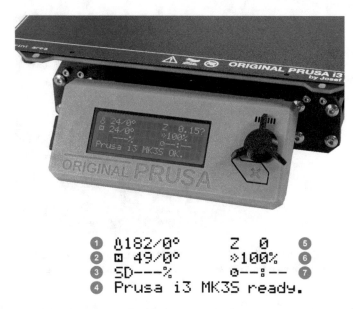

Fig. 9.1 The Prusa i3 MK3S+LCD panel. © Prusa Research—prusa3d.com

9.2.3 Loading Filament

Filament is mounted on a sturdy spool holder above the printer and loaded up and over the spool, from back to front. There are several diameters of filament used by different 3D printers. For example, Prusa printers use 1.75 mm diameter filament; other printers such as Ultimaker and LulzBot machines use 2.85 mm diameter filament. Make sure to purchase the filament diameter that works for your specific machine.

If you live in a humid area, once filament has been opened, it's a good idea to store your spools in dry or atmospherically-controlled boxes. Many FDM materials can take on atmospheric moisture, which causes them to become brittle or difficult to print. If your filament is making lots of snap, crackle, and popping noises like a bowl of Rice Crispies, it has likely taken on too much moisture and will need to be dried out to guarantee an acceptable print quality.

Once you've selected a filament to use, rest the filament spool on top of the filament holder. We'll load it once the printer heats up. Always make sure you're either holding the filament spool tightly or that the filament is secured through a series of holes or notches located on the filament spool itself. Only take the filament out of the secure filament spool holes when the printer is heated up and you're ready to load it. Accidentally letting filament unspool or unwind can cause tangles, jams, and print failures.

It's also a good habit to clip the end of the filament at an angle, both before you start if it's not already clipped and after your finish and unload your filament again (Fig. 9.2).

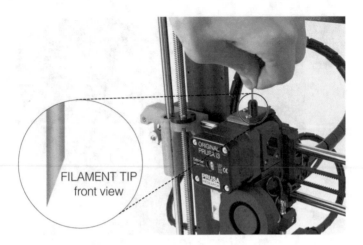

Fig. 9.2 When loading material, it's a good habit to clip the tip of your filament at an angle to prevent clogs. Image from the Prusa i3 MK3S+ manual. © Prusa Research—prusa3d.com

While the Prusa heats up, inspect your filament. Ensure there is no blob or blob at the tip of the filament. Use clippers to remove any blobs and create a nice, angled point in its place.

Once the machine is at temperature, select the *Load Filament* option from the main menu, insert the filament into the hole at the top of the extruder, and push down with light but steady pressure. If the printer is actively moving the hobbed bolts to load filament, you should hear a faint noise and feel the hobbed bolts gently grip and pull the filament. Once you feel the machine grip the filament, release your grasp and wait until filament extrudes from the nozzle in the correct color. The printer may pause after a period and ask if the filament is the correct color. Continue extruding filament until the proper color is coming out unless otherwise desired. Discard or recycle the ejected material.

Another important concept to note here is temperature. Typically, it's good practice only to have the printer sit at material melting temperatures when changing filament or actively printing. Recall from our discussion of the concept of heat creep that it's not a good idea to have the printer sitting idle at temperature for too long. In all cases, filament should be moving through the hot end at a steady rate. If the printer is stopped and not printing (or not getting ready to print), the machine should not be at temperature. If it is, it can cause heat creep and result in a jammed extruder.

9.2.4 Starting Your Print

Once your filament is loaded, navigate to your project from the LCD main menu by selecting *Print from SD* and locating the file. Before printing, be sure to inspect the printer nozzle and build plate and remove any debris from the build plate, like old primer lines (which we explain below). These can be thrown out or recycled. Just like with your slicer settings, it's a good idea to run through a quick mental checklist for the machine itself to ensure that you've done everything correctly before hitting *Print*.

- Is the build plate clear? Is the previous primer line or primer blob removed and thrown out or recycled?
- Is the correct filament loaded? Is the color extruding correctly?
- Is the nozzle clean? It's important to ensure no residual gunk is stuck to it or the surrounding hot end. If there is, you should cancel your print by hitting the *X* button below the LCD knob. Then, preheat the printer for the material you're using. When the printer is at temperature, use a brass brush to very carefully and gently brush the nozzle to remove gunk build up. Avoid touching the brass brush to the fragile thermistor wires, which are usually red and white and coming out of the heat block.

Then, the exciting part: simply select your project and press the LCD knob to begin 3D printing.

The printer may take a few minutes to heat up. Once it does, the printer will perform a series of nine short probes around the build plate using a special sensor (Fig. 9.3). This PINDA probe knows that it should be a set distance away from the build plate at each of these nine points. Therefore, it can detect if the build plate is higher or lower than expected, which ultimately determines if the build plate is level enough to print.

Fig. 9.3 The PINDA probe, used to detect if the build plate is level. Many modern printers have similar bed leveling technologies. Image from the Prusa i3 MK3S+ manual. © Prusa Research—pru sa3d.com

Once the probe has finished checking the build plate, the 3D printer will begin to print. A primer line is usually extruded at the front of the build plate before printing begins. This ensures that the nozzle has filament of the correct color and that the printer is laying down filament correctly. Primer lines, primer blobs, and bed adhesion skirts also exist to ensure that filament is at the very tip of the nozzle, ready to extrude. Though they're very clever machines, 3D printers aren't quite clever enough to tell if they have thermoplastic sitting at the very end of the nozzle or if some of the filament has seeped or oozed out over a short period of time, leaving an air gap inside the nozzle's tiny reservoir. If you were to print with an air bubble in the nozzle, the printer would simply move as if it were printing normally, leaving a gap or hole in your model and often causing the print to fail. The easiest way to fix this? Simply extrude some filament in a primer line, primer blob, or skirt right before the print itself to ensure that the printer is primed and ready to go. Most 3D printers automate this process, but you can add or remove primer lines and blobs in *Custom* settings of Cura [2].

It's also a good rule of thumb to supervise the first full ten minutes of any print. Why? To make sure everything is laying down and that material is extruding properly. Remember, a 3D printed part is like a house: it can't stand without a sturdy, well-laid foundation. Then, if everything looks like it's laying down well, you should be in the clear. Periodically check your project to ensure that the project is printing as intended. Congrats! You've just successfully started your first 3D print. Read on to learn what to do once your print has finished.

9.2.5 Unloading and Storing Filament

Since it's best to keep unused filament in dry storage, you should unload filament once the machine is finished printing. First, select the *Unload Filament* option from the menu. If the machine has cooled down significantly beforehand, you may need to preheat it again by selecting your material. Once ready to unload filament, the machine will beep and prompt you to click the LCD knob. It will then rapidly reverse the material out of the top of the extruder. Make sure to hold the filament and remove it from the extruder, grasping it firmly to ensure the filament doesn't accidentally unspool. If there is a blob or bulb at the end of the filament from where it began to melt while printing, use wire cutters to remove the blob and trim the end to an angle. Then firmly secure the end of the filament to the spool through the holes on the sides of the spool to prevent it from unwinding before placing it back in your storage box.

Most 3D printers like the Prusa are perfectly fine to sit on and idle without using too much power. However, it is important to make sure that it is not sitting at temperature for long periods of time. You can press the X button under the LCD knob to reset the printer back to its ready state, which also resets and turns off the heating elements. To fully turn off the Prusa, locate the power switch at the back of the machine near the power transformer and switch it to the OFF position.

Once the machine has cooled down, don't forget to clean up any debris from on or around the build plate. This includes things like discarded supports or primer lines. If the hot end needs to be cleaned, it must be cleaned at temperature with a wire brush. Try your best to avoid touching the build plate with your bare hands, as the oils within your hands may affect the adherence of prints to the build plate.

If you need to cancel your print or if you suspect the machine is malfunctioning in the middle of a project, quickly cancel your print by hitting X or turning off the machine and then remove your incomplete project. Take some time to service the machine or clean specific components before you restart your project. Simple malfunctions might include scary noises such as the 3D printer stepper motors grinding on an axis, the nozzle coming into direct contact with the build plate, the 3D printer making repeated beeps or unexpected electronic noises, or filament failing to extrude correctly. In cases of major mechanical failure, you should immediately shut off the machine using the OFF switch before diagnosing the problem to prevent further damage.

9.3 Best Practices

- After slicing and before exporting to an SD card or USB, always:
  - Inspect your model via *Preview* from top to bottom, and zoom in to ensure small features will print correctly.
  - Ensure that the first layers of your project are effectively touching the build plate.
- Every material has different properties when printing. Be sure to consult the manufacturer's recommendations for printing parameters like temperatures and speeds if you are unsure how to use the material. Always work in a well-ventilated area, even if the material you are printing with does not create fumes.
- When removing supports from FDM 3D prints, excess material can fly off in tiny shards. It's good practice to wear safety glasses to prevent eye damage.
- FDM 3D printers must have filament loaded for the entirety of the print, or else the project will fail. Therefore, always make sure you have enough filament for the whole project.
- Many printers are limited by build volume size. If you wish to print objects larger than the build dimensions, you may need to split your project into pieces first.
- Many FDM 3D printers have built-in safety features such as emergency stops and power loss protocols. However, this does not mean that they can be left unattended for large amounts of time. Make a habit of checking on your active projects periodically.
- Importantly, you should always supervise the first ten minutes of any print to ensure the project has a strong foundation.

9.4   Summary

This chapter detailed the last steps necessary before hitting *Print* on the 3D printer. First, we discussed a few last key settings to double-check before exporting the G-code file. With our fictional SD card in hand and ready to be brought over to the Prusa i3 MK3S+ 3D printer, we explored how to navigate the Prusa LCD menu and interpret the information on the home screen. Next, we reviewed the proper ways to handle and load filament, and general best practices for storing filament. We discussed how to load troublesome filament and best practices when the printer is on and sitting at temperature. Next, we learned how to start a project, and what things to watch for in the first ten minutes of any project. Finally, we wrapped up the chapter with a discussion on how to unload and store filament.

9.5   Chapter Problems

- Why is it good practice to enter *Preview* mode and scroll down to the first few layers of your sliced project?
- Where can you see the remaining project print time on the Prusa LCD menu?
- What diameter filament do Prusa i3 MK3S+ printers use?
- What can happen with filament spools in humid areas?
- What purpose does clipping filament at an angle before printing serve?
- Should you leave your printer on at temperature when you're not printing?
- What purpose do primer lines serve?
- What does the PINDA probe on a Prusa i3 MK3s+ do once you hit print on the navigation menu?
- Why is it a good habit to supervise the first ten minutes of any print?
- What's the typical procedure for unloading and storing filament?

References

1. Prusa Knowledge Base web page. https://help.prusa3d.com/.
2. Ultimaker. Ultimaker Support web page. https://support.ultimaker.com/.

'Steam-Building' Exercises

10

10.1 Objectives

Objectives: Let's put it into practice. The final chapter will highlight a number of ways to integrate 3D printing across a range of different science, technology, engineering, art, and math ("**STEAM**") projects. We'll present you with the information and resources to explore and potentially build the project yourself, as well as some background information to provide context to these specific examples. This chapter will review 'STEAM-building' projects that include:

- Paleontological puzzles, and how you might use 3D printed fossils as teaching and research tools in science fields
- Prototyping prosthetics: building your own DIY prosthetic hand
- Creating, customizing, and coding a 3D printed Arduino robot
- Making music with a variety of 3D printed instruments
- Visualizing complex math concepts with 3D printed teaching examples

10.2 Science Application: Digitizing Fossils

You might think of paleontology and archaeology as somewhat dusty science fields. Maybe you picture rows and rows of bones and artifacts stacked inside dim, dark wooden collection halls within old Victorian buildings or museums. You'd be wrong! Often, science fields such as paleontology (the study of fossils), anthropology (the study of humans), and archaeology (the study of human artifacts and material remains, a sub-field

© The Author(s), under exclusive license to Springer Nature Switzerland AG 2022
T. Kerr, *3D Printing*, Synthesis Lectures on Digital Circuits & Systems,
https://doi.org/10.1007/978-3-031-19350-7_10

of anthropology) are some of the earliest adopters of emergent technology like 3D printing [1–5].

Today, 3D printing is used by paleontologists in a huge number of ways, and is often complemented by another technology: 3D scanning. Scientists can use 3D printing to create casts and even photorealistic models of fossil displays, where the original is perhaps too precious, rare, or fragile to be publicly displayed. Scientists can use 3D scanners to capture incredibly small and even internal 3D models of fossils. They can create digital and physical copies of fossils that no longer exist—maybe the fossil naturally weathered or broke into pieces over time, or was broken or ground up intentionally to conduct geochemical analyses. Scientists can scan microscopic fossils and scale them up to place them in the hands of instructors, researchers, and students. They can capture 3D scans of fossils in rock outside in the field, and take those scans back to the lab to digitally prepare, excavate, and even assemble fossils without ever needing to touch the original rock. These assembled, articulated 3D models can then be 3D printed, painted, and put on display to make lightweight, accurate casts to show museum visitors and even researchers. Because of their ease of access and reproducibility, 3D printed casts are taking over for traditional heavy plaster casts and replicas. 3D prints can even be used to digitally remodel, repair, and patch up fossils that are cracked, broken, or have big missing pieces (Fig. 10.1). Most fossil skeletons you see on display in museums such as the Naturalis Biodiversity Center in Leiden, Netherlands are made up of real fossils interspersed with casts that are masterfully painted to look like the real thing [6]. Here's a fun challenge: next time you're in a natural history museum, look up at one of the giant, long-necked sauropod skeletons. Can you tell the difference between real fossils and casts?

This sounds exciting, but you might be curious how it may affect you. The answer? It's all about access! More and more scientists and museums all over the world are making their 3D scanned fossils available to download for free to anyone with a computer or 3D printer at home. Anyone can now access vast fossil collections, and the number of institutions digitizing their collections is growing by the day. Today, you can visit https://sketch fab.com/ to download an extraordinary assortment of fossils from academic institutions all around the world, such as the University of Wyoming, the Natural History Museum in London, the Smithsonian, the University of Dundee, the British Museum, the Idaho Virtualization Laboratory, and often directly from individual scientific researchers themselves. You can visit digital collections from the Smithsonian Museum's website (https://3d.si. edu/) or other smaller institutions (https://www.morphosource.org/) to download fossils, priceless artifacts, cultural heritage collections, and famous artwork. You can even visit https://africanfossils.org/ to download any of the fossils from the world-famous Olduvai Gorge in Tanzania, where the Leakey family discovered some of our earliest ancestors! 3D printing has a vast range of applications when it comes to scientific research, education, and informal outreach.

Let's put what we know about 3D printing and learn a bit about jaws, diet, and how much teeth can tell you about animals—even animals that are long extinct! First, we

Fig. 10.1 Dutch museum
Naturalis uses large 3D
printers to partially restore
their dinosaur fossils, such as
this *Triceratops* on display
(top). Real fossil (bottom, in
red) can be distinguished from
3D printed casts (white). 3D
model by Pasha van Bijlert

should rewind the clock a few million years to get some context. We know that a massive
mile-wide asteroid crashed into the Yucatán Peninsula along the southern border of the
Gulf of Mexico around 66 million years ago. With it, three-quarters of the plant and animal life on earth was wiped out in a catastrophic global extinction event. But surprisingly,
the dinosaurs didn't go extinct! While most large dinosaurs, such as *Triceratops, Tyran-
nosaurus, Hadrosaurus,* and *Ankylosaurus,* died out, some small avian dinosaurs survived.
These smaller ground and water fowl eventually radiated out and evolved into all modern
species of bird that we see today. This means that you can *technically* adopt dinosaurs
as cute talkative pets, order fried dinosaur at McDonald's, or enjoy delicious dinosaur at
Thanksgiving.

10.2.1   A Paleontological Puzzle: Whale Diets Through Time

The extinction of most large plant and animal life left large empty gaps in many different
ecological niches—gaps that a wide variety of smaller surviving species started to take
advantage of and fill. One amazing example is the evolution of whales. We know that
whales such as the blue whale are the largest animals to ever have existed on earth—even
larger than any of the known dinosaurs. But whales weren't always so big. Believe it
or not, the earliest whales we know about started off as small, hooved, mousedeer-like

Fig. 10.2 A lesser mousedeer from the Singapore Zoo. Photo by Bjørn Christian Tørrissen, CC-BY-SA-3.0

omnivores similar to mousedeer today (Fig. 10.2) who foraged, hunted, and waded along riverbanks 48 million years ago.

With giant swimming reptiles and large dinosaurs extinct, these early whales, or "archaeocetes," discovered that they could easily take advantage of an abundance of fish and small marine life that nothing else was eating. Over time, archaeocetes spent more and more time in rivers, then bays, and ultimately ventured out into the open ocean. As they did, they grew more and more aquatic, swapping fur for blubber and hooves for flippers. At the same time, they swapped jagged, critter-chomping molars, premolars, canines, and incisors for long uniform rows of pointed, conical, fish- and squid-spearing canine teeth. In some cases, whales even lost teeth entirely in favor of baleen for filter-feeding on huge shoals of krill. Over the span of only around 15 million years, whales went from small wolf-like river and coastal predators (*Pakicetus*) to furry shallow water, crocodile-like ambush predators (*Ambulocetus*) to streamlined, fully-aquatic, serpent-like hunters (*Basilosaurus*) to the whales we know today like *Orcinus orca*, the killer whale (Fig. 10.3).

Fig. 10.3 We'll use the incredible story of whale evolution as a case study of how 3D printing can be used to conduct comparative anatomy research

Now, let's download three digital models and 3D print them to see what their teeth can tell us about diet and behavior from 48 million years ago to the present. We won't coach you on 3D printing methods here. Instead, use what you've learned about 3D printing throughout this book to make sure you've selected the correct settings for your 3D printer. If your project doesn't come out the way you'd like, refer to *Sect.* 8.4: *Troubleshooting Common Issues* to try again. When you're ready to begin, download and 3D print:

- A bottlenose dolphin skull 3D scanned by Lily Wilson (CC BY 4.0) from the Anatomical Museum at the University of Edinburgh. https://skfb.ly/6SVAt
- A fossil *Coronodon havensteini* skull produced by Dr. Morgan Churchill and the College of Charleston's Mace Brown Museum of Natural History. Surprisingly, *Coronodon* is an ancestor of the mysticetes, or modern baleen whales. Notably, modern whales don't have any teeth at all. https://skfb.ly/6SsT9
- A leopard seal skull 3D scanned by Susanna Brighouse (CC BY-NC-SA 4.0) from the D'Arcy Thompson Zoology Museum at the University of Dundee. https://skfb.ly/6BFPu

Compare your 3D printed skulls to the five pictures (and linked resources) below and see if you can draw some conclusions about the function of the teeth and how they relate to diet and behavior.

Pakicetus (Fig. 10.4) is one of the earliest archaeocetes, or ancestral whales. Scientists know from both anatomical studies and geochemical analysis that *Pakicetus* was more of a dietary generalist compared to modern whales. It likely lived along river channels, occasionally submerging and diving after fish, while also hunting small animals on land. The wide range of teeth it had—from distinctly different molars, premolars, canines, and incisors—tells us that *Pakicetus* chomped, ground up, and chewed its food.

What similarities and differences in diet can you infer between *Pakicetus* and the 3D prints you've created?

Bottlenose dolphins (Fig. 10.5) are a modern species of oceanic dolphin that typically hunt fish, but also eat shrimp, squid, and other small ocean critters. They don't use their teeth to chew, but rather to catch and pin their prey before gulping them down whole.

Fig. 10.4 Cast of a *Pakicetus* fossil at the Canadian Museum of Nature. Photo by Kevin Guertin (CC BY-SA 2.0)

Fig. 10.5 An Atlantic bottlenose dolphin skull. Photo by James St. John (CC BY 2.0)

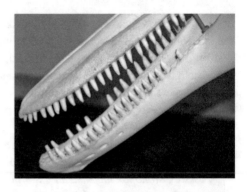

Fig. 10.6 A Leopard seal skull. Photo by Nkansah Rexford (CC-BY-3.0)

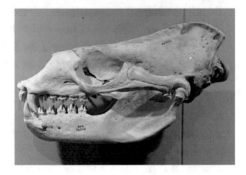

Do bottlenose dolphins have any sharp blade-like teeth for slicing, chewing, and cutting up prey? What about flat molars for grinding up food?

Leopard seals (Fig. 10.6) are another modern species. They are fierce, active predators that hunt everything from other seals to fish, birds, krill, and penguins. Leopard seals have sharp teeth to spear fish and birds, but also a jaw with gaps between premolars and molars and unusual-shaped teeth that can interlock to act as a strainer to catch krill.

Can you draw any parallels between your 3D printed leopard seal and *Coronodon* skulls? For example, how do the molar, premolars, canines, and incisors of *Coronodon* compare to leopard seal jaws and teeth?

Believe it or not, crabeater seals (Fig. 10.7) don't actually eat crab. Instead, they have funky, swirling teeth with special gaps that allow them to filter-feed on krill in a similar manner to baleen. You can explore a 3D model of a crabeater seal skull made by the Evans EvoMorph Lab at Monash University here: https://skfb.ly/6Uz9T.

Does the *Coronodon* skull have similar teeth? How do its teeth differ?

Mysticetes like gray whales (Fig. 10.8) don't have any teeth at all. Instead, they use huge bristly comb-like baleen plates to filter-feed tiny ocean organisms. Interestingly, many fossil baleen whales like *Coronodon* clearly have mouths full of varying types of teeth, which tells us that mysticete whales definitely had toothed ancestors.

Fig. 10.7 A crabeater seal
skull and teeth. Image
modified from original by
Nkansah Rexford (CC-BY-3.0)

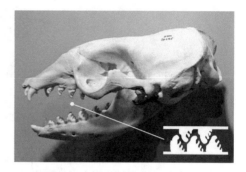

Fig. 10.8 A gray whale skull
with baleen attached. Photo by
Mira Mechtley (CC BY 2.0)

Can you make some inferences on how baleen whales slowly swapped out teeth for baleen as they evolved? Do you think *Coronodon* could have been a filter-feeder? What anatomical similarities does *Coronodon* have with another animal on this list?

Answer: Did you figure out how *Coronodon* may have dined on sea life? It was an early ancestor of modern baleen whales, which don't have teeth. We can observe that *Coronodon* has many teeth, but those teeth are spaced relatively far apart and are jagged in appearance, which mimic both leopard seals and crabeater seals. Scientists currently think that *Coronodon* used their unique teeth to filter-feed in a similar manner to crabeater seals do today!

10.2.2 Comparing Cats and Dogs

Want another science experiment? Let's make some scientific deductions to determine what each of these mystery animals are and what they may have eaten (Fig. 10.9). Download and 3D print the following 3D models:

Fig. 10.9 Scientists can use 3D printed skulls, such as from bobcat, coyotes, hyenas and racoons, to conduct morphological assessments. Importantly, the images above are not in order and not to scale! 3D models from the University of Wyoming Museum of Vertebrates collection (CC BY 4.0)

- *Procyon lotor*, scans made by the University of Wyoming Libraries from the UW Museum of Vertebrates collection (UWYMV, CC BY 4.0). https://skfb.ly/6VBAo. You can download additional files by visiting https://arctos.database.museum/guid/UWYMV:Mamm:5429.
- *Canis latrans*, scans made by the University of Wyoming Libraries from the UW Museum of Vertebrates collection (UWYMV, CC BY 4.0). https://skfb.ly/6XtBx. Visit https://arctos.database.museum/guid/UWYMV:Mamm:2860 to learn more.
- *Crocuta crocuta*, scans made by the University of Wyoming Libraries from the UW Museum of Vertebrates collection (UWYMV, CC BY 4.0). https://skfb.ly/6X9PD. Visit https://arctos.database.museum/guid/UWYMV:Mamm:5840 to learn more.
- *Lynx rufus*, scans made by the University of Wyoming Libraries from the UW Museum of Vertebrates collection (UWYMV, CC BY 4.0). https://skfb.ly/6YFzz. Visit https://arctos.database.museum/guid/UWYMV:Mamm:1464 to learn more.

Don't cheat and look up the identities of those skulls! Your challenge is to 3D print these all at roughly the same size, compare their teeth with a few helpful hints, and determine the identity of the mystery mammals using the clues below:

- One of these skulls is a bobcat. It has very sharp molars best suited for cutting, shearing, and slicing flesh rather than grinding it. Notably, cats have fewer teeth than dogs.
- One of these is a coyote. Its teeth are more generalized compared to cats, and are suited for slicing meat, grinding vegetable matter, and sometimes fruit. Notably, dogs have slightly more variation in teeth than cats.
- One of these is a hyena, which acts like a dog, but is more closely related to cats. It needs strong teeth built for crushing bone.
- One of these is a raccoon, which is more closely related to bears than to cats or dogs. Raccoons are some of the most omnivorous animals, and eat everything from insects, crayfish, worms, fruit, nuts, fish, amphibians, and sometimes even birds and small

mammals. Since they have such a varied diet, raccoons have the most variation in teeth: flat molars that grind and crush, premolars that slice, canines that grip and tear, and incisors that bite and cut.

Answer: Did you figure out what animal each skull belongs to? *Procyon lotor* is the common raccoon, which might help explain the wide variety of teeth. *Canis latrans* is a coyote, which eats mostly meat but also occasionally eats vegetables or fruit. *Crocuta crocuta* is a spotted (or laughing) hyena, sporting massive teeth to crush bone. Last, *Lynx rufus* is a bobcat, which almost exclusively eats meat.

10.2.3 The True Size of a Megalodon

Finally, 3D printing can be applied to paleontological studies where all you have are the teeth of an extinct animal! *Megalodon* was an extinct species of shark that lived approximately 23–3.6 million years ago, and grew to absolutely enormous sizes. Unfortunately for paleontologists, sharks don't have bones and therefore don't usually fossilize well. Instead of bones, they have skeletons made primarily of cartilage, a rubber-like tissue. While their skeletons don't usually preserve well during fossilization, their teeth, which are made of hard tissue, sure do! And because sharks produce many, many teeth in their lifetime, shark teeth are abundant in the fossil record. It's common to find both modern and fossil shark teeth washed up on the shoreline.

Your challenge is to 3D print a tooth from a full-size megalodon (*Otodus megalodon*) at the correct scale, and make some simple measurements from the start of the tooth's enamel to the tip of the tooth to estimate the full size of the shark [7]. This is a similar method of assessing the size of an unknown animal that scientists use, though we're using our 3D printing skills to create a life-sized tooth at the same time. Let's start by 3D printing a megalodon tooth from the Digital Atlas of Ancient Life (https://skfb.ly/6AFxu, fossil from the Paleontological Research Institution, by Emily Hauf, CC0 1.0). The creators of the megalodon 3D model state that the longest dimension is 16 cm, or approximately 6.3 inches. Since Ultimaker Cura works in metric, we'll want to make sure that the long axis of the tooth (its height, along the Z axis) is set to 160 mm. When printing, consider printing upright to minimize the amount of supports (and thus minimizing the amount of support scarring).

Once 3D printed, the tooth should be ready to measure (Fig. 10.10). Measure the enamel height of the shark tooth (H). In the original specimen, the enamel is the shiny part with small serrated edges. Write down the value you observed. We'll use this to get a very rough approximation of the shark's length in feet.

Fig. 10.10 Scientists can use
3D printed teeth to interpret
the size of enormous sea
creatures. Original image by
Francesco Volpi Ghirardini
(CC BY-SA 4.0)

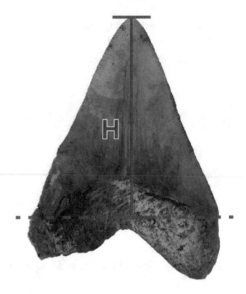

If you measured in inches, simply multiply the enamel height (H) using the following
equation to get your shark's approximate length in feet.

$$H \times 10 = \underline{\quad\quad} \text{ feet long}$$

If you measured the enamel height (H) in centimeters, use the following equation:

$$(H \div 254) \times 10 = \underline{\quad\quad} \text{ feet long}$$

Answer: If you 3D printed effectively and measured correctly, you likely got a
value around 40 or 50 feet long! Scientists believe that megalodon sharks grew as
large as 66 feet, and weighed as much as 228,000 pounds. It's more likely that most
megalodon sharks were around 30–35 feet long, which means that our megalodon
shark was just a bit above average when they were swimming around North Carolina
during the Miocene. It's worth noting that these results could be a product of the
very simplified math we conducted, which isn't as precise as what scientific studies
may use.

Were you surprised by these findings? Are you glad that Megalodon went extinct around
3.6 million years ago? As a fun final wrap-up fact, scientists have found evidence of deep
gashes on large whales and lots of megalodon teeth lying near gnawed-on whale fossils,
meaning that megalodon likely hunted *whales*.

10.3   Technology Application: Prototyping a Prosthetic Hand

As we've learned throughout this book, 3D printing is not a technology exclusive to engineering fields. Additive manufacturing can be used across a huge, diverse range of industries and a near-limitless number and type of projects. One major sector that's making heavy use of 3D printing is the healthcare industry. From anatomical teaching models to patient-specific replacement parts like knees, joints, and dental crowns, or from machine parts such as respirator valves to functional organs like *beating human hearts*, the opportunities are boundless. Medical fields around the world are taking advantage of tons of 3D printing to explore the future of healthcare as we know it. While some 3D printing applications require special labs, others can be made with your 3D printer at home. One particular application seems almost form-fitted for desktop 3D printing, and that is prosthetics [8–10]. Later in this 'STEAM-building' exercise, we'll cover how to make a prosthetic arm, but first, we should cover why accessible 3D printing is so important to the prosthetic industry in a bit more detail than we did in Chap. 2. Importantly:

1. **3D printing makes prosthetics more affordable and accessible.** Manufacturing custom prosthetics can get expensive very quickly. Today, the ability to 3D print prosthetics allows anyone to create parts using inexpensive 3D printing filament for thousands of dollars less than industry-manufactured prosthetics. Moreover, 3D printing makes it equally possible for someone with a $300 3D printer to create prosthetics that were previously only accessible to manufacturing groups that could afford and house expensive equipment, such as CNC mills or injection molding machines. It also opens the door for folks in rural and remote communities who may not have access to the resources, equipment, and facilities that larger population centers have to create their own custom prosthetics for members of their communities.
2. **It allows people at home to innovate, iterate, and customize.** Give everyone the means to rapid prototype and iterate ideas, and they'll be able to create more innovative products quickly! Better still, you can create adapted devices directly designed for a specific patient like a friend or family member who may have unique requirements for a proper fit. This increased level of customization certainly benefits the patient and can lead to some quality of life improvements.
3. **It encourages innovation and new technology discoveries.** When folks from all different backgrounds and perspectives approach a problem together, it can lead to all sorts of new ideas and technological advancements. Simply stated: diverse, inclusive communities are the engines of innovation. Thus, when students, researchers, and hobbyists at all levels have equal access and opportunity to innovate with technology such as 3D printers, they often create new ways to address challenges. This, in turn, leads to more growth across industries.

So what's the future of 3D printed prosthetics? It's a bright and exciting one, with tons of different applications that address cost, customization, and comfort. Let's learn how to 3D print our own prosthetic hand using a fantastic model designed by skilled teams at e-NABLE [11]. The e-NABLE Community is a worldwide volunteer group of 40,000 + 3D modelers, 3D printers, engineers, scientists, medical professionals, makers, and innovators. They have united to help design and print custom, free, open-source upper limb prosthetics for folks who may have been born missing fingers or hands or lost limbs due to war, disease, or natural disaster.

10.3.1 Downloading the Prosthetic Files

With such a large creative volunteer community, there are many different e-NABLE files freely available to download, including plenty of custom-made prosthetics. Of course, you can always learn more at https://enablingthefuture.org/ or download unique 3D files by visiting the e-NABLE resource hub (https://hub.e-nable.org/s/e-nable-devices/wiki/ove rview/list-categories), but for now, we'll focus on a simple design. You can download a 3D printable Phoenix 3 e-NABLE prosthetic hand by visiting: https://www.thingiverse. com/thing:4056253. Notably, the Phoenix 3 hand is wrist-powered, requiring a functional wrist to rotate the prosthetic.

The Phoenix 3 prosthetic (Fig. 10.11) was skillfully designed by Jason Bryant, John Diamond, Scott Darrow, Andreas Bastian, Team Unlimbited, e-NABLE France, and Jeremy Simon.

Fig. 10.11 The Phoenix 3
prosthetic hand, created by
talented designers at e-NABLE
(CC BY 4.0)

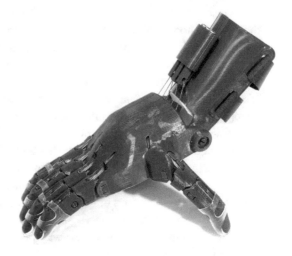

10.3.2 3D Printing Instructions

Ready to 3D print the prosthetic Phoenix 3 prototype? The good news is that the teams at e-NABLE have put lots of thought and care into making the 3D printing process as easy as possible. You won't need any type of supports, nor any type of raft. If you were to print with a brim or skirt, and if you simply set your infill set to 30–40% (to ensure that the parts are strong), that's all you'll need to do! Refer to the device catalog instructions at https://hub.e-nable.org/s/e-nable-devices/wiki/e-NABLE+Phoenix+Hand+v3 to get started.

A few things worth noting:

- e-NABLE created files for both a left and a right-handed model.
- e-NABLE has also provided labels for each of the small pins (Fig. 10.12) that can help you identify pin locations.
- If you're making this for a specific person, you may need to scale up the size of all files. At 100% scale, the prosthetic is fit for a smaller child. Make sure to measure the person you're making the device for by using the measurement guide found at https://hub.e-nable.org/content/perma?id=391.

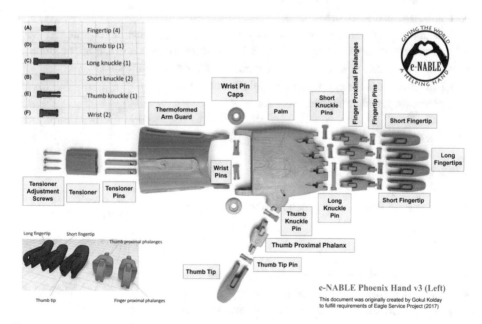

Fig. 10.12 Instructional diagram for the e-NABLE Phoenix Hand v3, created by Gokul Kolday (CC BY 4.0)

10.3.3 Additional Material

While 3D printing should be straightforward because of how much time the e-NABLE team invested in making sure the models were easily printable, the assembly of the prosthetic will take more time. It will also require that you source some parts, such as screws, metal cord, elastic, and foam padding. You can purchase an inexpensive kit on the e-NABLE website by visiting https://shop3duniverse.com/products/phoenix-hand-by-e-nable-assembly-materials-kit. Alternatively, you can source these parts yourself. In addition to PLA filament, you'll need some standard workshop tools and some parts purchased online, or from a hardware store:

Common workshop tools

- 1 pair of slip joint pliers
- 2 pairs of long-nose pliers
- 1 small round file to file down any joint pin holes
- 1 medium half-round file to help to trim down excess foam
- 1 small flat file to (optionally) file or sand down pin heads
- 1 small hammer
- 1 pair of wire cutters to cut the "tendon line"
- 1 Phillips-Head screwdriver
- 1 medium-sized clamp to hold the hinge when stringing up the finger joints
- 2 wooden toothpicks to help you attach the small dental bands to the finger joints
- A 1/8″ pin punch or small nail to help push any tight pins into joints
- 1 pair of scissors for cutting card templates and foam
- 1 pencil for tracing templates onto foam

Tensioner Screws

- 4 × Pan head Phillips sheet metal screws #4 × 3/4″
- 4 × Pan head Phillips sheet metal screws #6 × 1″
- 4 × Pan head Phillips sheet metal screws #8 × 1 ¼″

Palm Screws

- 15 × Countersink head Phillips wood screws #4 × 3/8″
- 15 × Countersink head Phillips wood screws #6 x ½″
- 15 × Countersink head Phillips wood screws #8 × 5/8″

Cords ("Tendon" Lines)

- 16 feet of 80 lb strength braided fishing line

- 10 feet of Flexible cord 1.0 mm
- 10 feet of Flexible cord 2.0 mm

Elastic Bands

- 100 × Non-latex extra heavy grade dental bands 1/4″ size
- 100 × Non-latex extra heavy grade dental bands 5/16″ size

Firm Foam Padding

- 12″ × 6″ piece of 1/8″ thick self-adhesive firm foam padding

Gel Fingertip Grips

- 10 × Size 3 Lee Tippi Gel Fingertip Grips

Velcro Straps

- 2 × Velcro straps, 12″ long, 1″ wide with buckle

10.3.4 Assembly

Finally, you can follow very detailed assembly instructions created by e-NABLE volunteer John Diamond for the Phoenix prosthetic: https://cdn.thingiverse.com/assets/dd/6b/45/30/fc/Phoenix_v2_assembly_guide.pdf. He also produced a great video tutorial for the assembly process, which you can watch at https://youtu.be/Der_DD2_zps.

10.3.5 What's Next?

Great work! With one 3D printed prosthetic now assembled, think about how you might customize another 3D printed prosthetic. What would you change? What added functionality would you want to include in a Phoenix Prosthetic Hand v4? What about additional fingers, hands, toes, or feet?

Enterprising innovator Dani Clode has invented devices that can give anyone a *third* thumb, allowing people to tackle jobs that typically take two hands with only one! Believe it or not, the Third Thumb project is a 3D printed thumb extension that uses pressure sensors fitted into shoes to control motors in a thumb via Bluetooth simply by flexing your toes. And these designs aren't just engineering art: Dani's Third Thumb is currently being

Fig. 10.13 3D-printed Third Thumb invented by augmentation designer Dani Clode. Visit https://www.daniclodedesign.com to learn more about Dani's other body-bending designs

used in neuroscientific research in collaboration with The Plasticity Lab at University College London & Cambridge University to explore how humans might harness additional prosthetic tools to augment and extend their physical and cognitive abilities [12]. You can read more about her Third Thumb prosthetic project (Fig. 10.13) [13], as well as some other amazing projects, by visiting https://www.daniclodedesign.com/thethirdthumb.

10.4   Engineering Application: 3D Printed Arduino Robot

Interested in tying 3D printing into robotics and microcontrollers such as Arduino [14, 15]? Let's 3D print and build the popular, open-source "Otto" Arduino robot (Fig. 10.14) [16, 17]. Using your knowledge of 3D printing, you'll be able to create and even possibly customize the outer chassis of the robot, and eventually even use your smartphone to make Otto dance, walk, sing, and navigate around obstacles. You can buy Otto Maker kits online via https://www.ottodiy.com/store/products/otto-diy-starter and customize your own CAD designs for Otto via https://www.ottodiy.com/design.

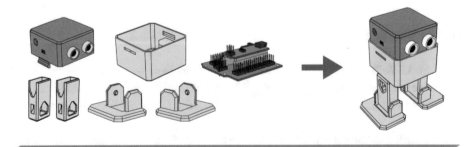

Fig. 10.14 3D printed Arduino "Otto" robot

10.4.1 3D Printing the Otto Robot .STL Files

Otto is available in two kits: a pre-made Builder ker kit, which only provides the electronic components. The Maker kit is for those readers who have access to a 3D printer and want to 3D print the parts themselves, which we'll want to do here. You can download the Maker kit 3D model files for free by visiting https://www.ottodiy.com/academy.

To start, download all the files for "Otto Starter" and load them into the Cura slicer (Fig. 10.15). For this example, you'll want to import the "new PCB" model files. If you've purchased a SparkFun Otto model from the Otto website, use those 3D files instead. Can you use what you've learned about 3D printing to address the following challenges?

- Otto's 3D file repository only comes with one leg.STL file. This is on purpose. Do you remember how to duplicate objects in Cura?
- There are many small parts to the Otto assembly. How can you rearrange the entire Cura build plate in one quick step to optimize the build plate for printing?
- When imported into Cura, Otto's head is oriented upside down. To save lots of time and support material, it may be best to flip the head 3D model upside down. How can you select a specific face and align it to the build plate?

Fig. 10.15 Once you've downloaded and imported the Otto 3D files, your build plate should look something like this. Screenshot from Ultimaker Cura (https://ultimaker.com) software

Since we're printing the parts ourselves, it's up to you to choose the parameters you'd like to 3D print with. Like the e-NABLE prosthetic prototype, Otto is well-suited to 3D printing and should 3D print very easily if you've followed this book closely. We recommended printing Otto using a Cartesian FDM 3D printer using the following settings:

- **Material:** PLA
- **Resolution (Layer Height) Profile:** 0.20 mm or 0.15 mm
- **Infill:** 20%
- **Supports:** None
- **Bed Adhesion:** None (which, if you remember, defaults to a skirt)

Once you've selected your settings, make sure to slice your project and head over to "Preview" mode to view your model and inspect the settings you've selected. Then, use the slider on the right side of the UI to scroll down to the base layers of your project and ensure that everything is laying flat, sufficiently contacting the build plate, and that your infill and support settings are accurate (Fig. 10.16). In our example, Otto should take around ten hours to 3D print and use about 96 g of material. Since filament spools are usually around 1000 g of material in total and typically cost between $20 and $30 per spool, we'll be able to 3D print plenty of inexpensive replacement or custom parts if we want! This means our Otto robot will only cost us around $2 to $3 in material.

Pop quiz: what happens if you start to change settings in Cura? Using some advanced slicer skills from Chap. 8, what settings might you change to shave off an hour or two and make the Otto project 3D print faster?

10.4.2 Assembling and Coding Otto

Otto is programmed to operate out of the box, once you've finished assembling the 3D printed parts. For up-to-date assembly instructions, visit https://www.ottodiy.com/academy. Once your 3D printed Otto is assembled, you can access the Otto Arduino IDE guide to begin coding. Before you do, you'll need to download the Otto DIY libraries by visiting https://github.com/OttoDIY/OttoDIYLib/. You'll also need to download the Arduino IDE to start coding, which you can access by visiting https://www.arduino.cc/en/software. This book, "3D Printing: Introduction to Accessible, Affordable Desktop 3D Printing," builds from Arduino concepts established in the Synthesis Lectures on Digital Circuits and Systems book series, primarily *Arduino I: Getting Started* by S. Barrett through *Arduino IV: DIY Robots* by T. Kerr and S. Barrett. For further reading and a comprehensive overview of all things Arduino, we recommend diving into those books!

Next, open the Arduino IDE and navigate to *Sketch > Include Library > Add.ZIP Library*. Find the location where you saved the Otto DIY libraries and open the file. You'll receive a confirmation that the library has been installed. To double-check, you

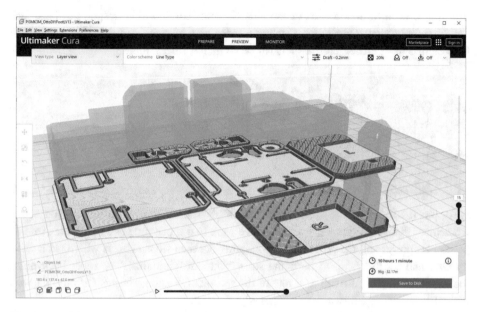

Fig. 10.16 Always make sure to check Preview mode and scroll down to the bottom of your sliced file in Layer view before pressing 'Print.' Can you recall why? Screenshot from Ultimaker Cura (https://ultimaker.com) software

can navigate to *Sketch > Include Library* menu. You should now see the "Otto DIYLib" library at the bottom of the *Include Library* menu. You can also search for and install the Otto libraries via *Sketch > Manage Libraries*. If you're curious, you can poke around some of the base libraries hidden within the primary Otto.h library:

- Otto.h and Otto.cpp contain all the primary functions.
- Otto_gestures houses all the gesture functions.
- Otto_mouths houses all the mouth functions.
- Otto_sounds houses all the sound functions.
- Otto_matrix houses all the matrix functions.

To add the primary Otto library to your Arduino project, enter:

```
#include <Otto.h>
Otto Otto;
```

You'll also need to define each of the pins on your Arduino to correspond to legs, feet, and buzzers on your Otto robot. Declaring these pins tells the Arduino IDE software to define pin 3 as "LeftLeg" for the left leg servo motor, for example.

```
#define LeftLeg 2 // left leg pin
#define RightLeg 3 // right leg pin
#define LeftFoot 4 // left foot pin
#define RightFoot 5 // right foot pin
#define Buzzer 13 //buzzer pin
```

Next, we'll need to write some code for Otto's startup procedure under void setup. Remember that "void setup()" is the code we want to run one time as soon as the Arduino board gets power and the program starts running but then stops. This is where we'll wake up Otto and set the robot to its home state. "Otto.init" initializes the robot by syncing the leg and feet pins, as well as a few of the sensors. Otto.home brings Otto to its home, neutral position: legs forward, feet oriented side-to-side.

```
void setup() {
    Otto.init(LeftLeg, RightLeg, LeftFoot, RightFoot, true, Buzzer);
    Otto.home();
}
```

Now the fun part. Otto has lots of customizable functions that you can change. For the sake of space, we won't detail them all here. Instead, we'll wrap up this section with a focus on a small subset of Otto commands that should allow you to see exactly how to start customizing code for your own personal Otto.

To import the example code, start a new project and navigate to *File* > *Examples* > *OttoDIYLib* > *Otto_allmoves*. There are tons of ways to customize your Otto, but we'll take a look at movement here. The servos and sensors in your assembled Otto work in tandem to let Otto shuffle forward and backward, side to side, jump, dance, sing, detect walls, and even moonwalk. You can change the values inside the parentheses for commands such as "Otto.walk(steps, T, dir)" to make Otto walk faster, slower, change directions, and even change the size or angle of the movement. In our Otto.walk example, parameters include the number of steps in the programmed time period, time (in milliseconds), and direction (1 for forward or left depending on the command, −1 for backward or right). For example, if we were to type "Otto.walk(5, 2000, −1)," our robot would walk backward five steps over a two-second (2000 ms) period. A higher time value means a slower movement. Try values between 500 and 3000 ms. Next, try to edit some of the following. What happens to your Otto robot? Can you predict how the Otto robot might behave just by reading the code?

- *Otto.walk(10, 1000, 1);*
- *Otto.turn(5, 500, −1);*. Recall that for turning, bending, and side-to-side shaking movement, 1 and −1 represent left and right.
- *Otto.bend(1, 2000, 1);*
- *Otto.shakeLeg(5, 3000, −1);*
- *Otto.jump(2, 500);* For this command, there is no direction of movement. Instead, Otto will simply flex its legs to stretch upward.
- *Otto.moonwalker(3, 1000, 25,1);* Note that there's one additional parameter here, between time (1000 ms) and direction of moonwalk (left, or 1). This represents a new parameter for all dance moves, "h" for the height or size of the movement. For the moonwalk, try changing the h value between 15 and 40. What happens?

A complete list of commands and their parameters can be found alongside the Otto library by visiting https://github.com/OttoDIY/OttoDIYLib/.

10.5   Art Application: Diy Musical Instruments

Many readers might have experience playing music, or may have always wanted to pick up and learn an instrument. For those who are musically inclined, consider making your own custom, inexpensive 3D printed instruments [18–20]! While 3D printed instruments may lack the beautiful resonance that wood or metal can provide, these are still fun projects to explore. In some cases, such as with many string instruments, you may need to source your own string, tuning pegs, and small fasteners or mechanical parts. In *all* cases, these 3D models will give you the freedom to 3D print your own unique instrument and customize it however you'd like! Use the guides below, and reference the recommended settings shared in each of these project profiles to make your own musical instruments.

10.5.1 Woodwinds

Some of the more accessible instruments to 3D print are woodwind instruments, particularly those that don't require any additional special parts, such as wooden reeds. As a starting point, reedless woodwind instruments such as flutes are a great place to begin your musical 3D printing journey.

Fig. 10.17 Recorder designed
by Joe Larson (CC BY-SA 3.0)

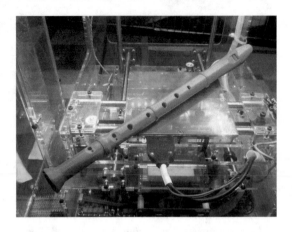

10.5.1.1 Recorder

Let's start off with something easy. You can 3D print your own recorder (Fig. 10.17) using "3D Printing Professor" Joe Larson's design by visiting https://www.thingiverse.com/thing:12301. Recorders are a type of woodwind instrument first documented in Medieval Europe. With no reeds required, recorders can create consistent music using only your breath and proper finger placement. Because of how easy they can be to pick and play, recorders are relatively common in schools.

How does this 3D printed recorder compare to one you might purchase in a music shop?

10.5.1.2 Ocarina

Ocarinas are vessel-shaped wind instruments often made of fired clay and sometimes made of wood, glass, bone, or even plastic. With a history dating back over 12,000 years in Mesoamerica and China, ocarinas are ancient instruments. In popular culture, you might have seen ocarinas played by Link, the hero of the Legend of Zelda video game series. Rob Beggs, a naval architect and navy technician, designed a trendy ocarina model (Fig. 10.18) that you can access by visiting the download page: https://www.thingiverse.com/thing:2755765.

10.5.1.3 Flute

Modern flutes are rooted in a vibrant history that goes back even farther than ocarinas to around 60,000 years ago, giving them the amazing distinction of being the world's oldest instruments! The oldest known flute was carved from the femur bone of a cave bear and discovered in a cave in northern Slovenia in 1995. More remarkable still, scientists believe Neanderthals made it! 3D model designer Paul Harrison from Melbourne, Australia made a 3D printable modern flute (Fig. 10.19) that you can make at home. As the designer notes, you can print your own flute in three different sizes: tenor in D, alto in G, or soprano in D. He was even generous enough to split the files into pieces that could fit on

Fig. 10.18 Ocarina created by Rob Beggs on Thingiverse (CC BY 4.0)

Fig. 10.19 Folk flute designed by Paul Harrison (CC0 1.0)

a wide range of 3D printers. You can access the files by visiting https://www.thingiverse.com/thing:162490.

10.5.2 Brass Instruments

Ready to toot your own horn? Try to print the following 3D models. Brass instruments that are 3D printed in plastic aren't likely to result in the beautiful music that the likes of Louis Armstrong, Miles Davis, or Alison Balsom would enjoy. Usually, the sound quality of a plastic trumpet will sound quite a bit different than an actual brass trumpet. Don't say we didn't warn you! These are still fun to 3D print and test out, however.

10.5.2.1 Trumpet
You're not likely to get the same bright, strong, brassy notes from a 3D printed trumpet that you would get with a traditional metal instrument. Still, you can consider mixing and matching an existing metal mouthpiece with your 3D printed trumpet (Fig. 10.20)

Fig. 10.20 Trumpet created
by Ozkal Ozsoy (CC BY-SA
3.0)

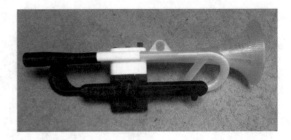

Fig. 10.21 Horn designed by
Robo on Thingiverse (CC BY
4.0)

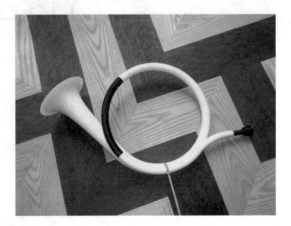

for better results. Inventor and electrical engineer Ozkal Ozsoy designed a 3D printable
trumpet that you can download by visiting https://www.thingiverse.com/thing:307088.

10.5.2.2 Horn

Prepare to be surprised: this horn can create unusually good music for a 3D printed
instrument. These simple horns are a type of valveless "natural horn" brass instruments
that were traditionally used by postilions to announce the arrival or departure of horses,
mail coaches, or post riders. User "Robo" designed a 3D printed horn (Fig. 10.21) that
you can download by visiting https://www.thingiverse.com/thing:121702.

10.5.2.3 Shofar

Shofars are an ancient type of musical horn that act a bit like a bugle, and are still made
of ram or kudu horn today. These horns are traditionally blown as part of Jewish religious
holidays, used to mark the end of Yom Kippur, during Rosh Hashanah services, and the
weekday mornings running up to Rosh Hashanah during the 12th month of the Hebrew
calendar, Elul. 3D printing company MakerBot designed a ram's horn shofar (Fig. 10.22)
that you can download by visiting https://www.thingiverse.com/thing:29975.

Fig. 10.22 Shofar designed
by MakerBot on Thingiverse
(CC BY 4.0)

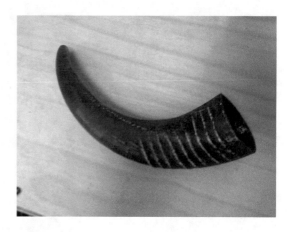

10.5.3   Percussion Instruments

Drum roll, please! While 3D printers aren't well suited for creating parts of percussion instruments such as the drum skin membrane, they are perfect for creating smaller hand-held instruments and percussion components. In fact, well-known bands such as *Panic! At The Disco* have even 3D printed their own drum designs: https://www.youtube.com/watch?v=_z7e5KmVAJs.

10.5.3.1   Frog-shaped Guiro

The *güiro* is a traditional percussion instrument popular throughout South America and Africa that's made from gourd-like fruit and played by scraping the bumpy ridges with a pick, stick, scraper, or tines. This frog-shaped guiro (Fig. 10.23) was created by Chandler Barr, and even comes with a customizable bowtie and hat. Barr recommends wrapping the stick in cloth or paper towel to make the best sound. You can visit https://www.thingiverse.com/thing:4525752 to download the model.

10.5.3.2   Maracas

If you're looking for a unique challenge, printing Pedro Moreira's 3D model of maracas (Fig. 10.24) requires that you pause the print around 30 or 40% to insert rice, beans, tiny pebbles, or other small things to make it rattle. Maracas and similar types of rattling percussion instruments originated thousands of years ago, and are historically found across a number of different cultures in the Americas, Pacific Islands, and Africa. You can download Moreira's maraca files at https://www.thingiverse.com/thing:938783.

Fig. 10.23 Frog-shaped güiro
designed by Chandler Barr
(CC BY 4.0)

Fig. 10.24 Maracas created
by Pedro Moreira (CC BY-SA
3.0)

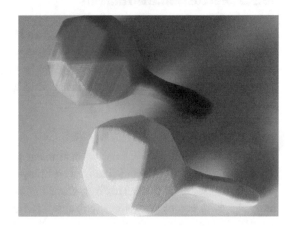

10.5.4 String Instruments

If you're up for the ultimate challenge, 3D printing string instruments might be right up
your alley. While these 3D prints require you to purchase additional parts such as strings,
tuning pegs, and fasteners, the payoff is perhaps more rewarding. 3D printed horns and
woodwind instruments might produce an off-key, out-of-pitch sound, but 3D printed string
instruments can more closely approximate the real thing.

10.5.4.1 "Banjolele" Banjo-Ukulele
There are tons of 3D models for small stringed instruments such as ukuleles available
online via Thingiverse.com or Printables.com, so we won't detail them all here. Instead,
we want to highlight a cool string instrument: the "Banjolele" designed by Andreas Bastian. Banjoleles were once quite popular in the 1920s and 1930s, but Bastian notes that

Fig. 10.25 Banjolele designed by Andreas Bastian (CC BY 4.0)

this banjo-ukulele 3D model (Fig. 10.25) was an idea born in 2013 after weeks of iterative design and collaboration with Geoff Wiley, co-founder of the Jalopy Theatre and School of Music in Brooklyn, New York. You can download the banjolele files for free by visiting https://www.thingiverse.com/thing:113908.

10.5.4.2 Modular Fiddle

"Fiddle" and "violin" are just two words for the same stringed instrument. The term "fiddle" is used more often when referencing folk music such as bluegrass. David Perry, founder and lead designer at OpenFab PDX, created a full-sized, modular fiddle (Fig. 10.26) that you can download and 3D print by visiting https://openfabpdx.com/fiddle/. If you're up for another challenge, try OpenFab PDX's fully functional electric fiddle that they titled the "F-F-Fiddle," a clever play on the term "fused filament fabrication." You can find the assembly instructions and download links on the OpenFab PDF website by visiting https://openfabpdx.com/fffiddle/.

10.6 Math Application: Visualizing Math

We'll wrap up this chapter with our final 'STEAM-building' exercises that might help students find fun and out-of-the-box ways to understand mathematical concepts. Math—particularly advanced or multivariable math—can sometimes be difficult to conceptualize. For many folks, complex math might be challenging to connect to the real world. For others, math textbooks might present information and illustrations in ways that are tough to grasp or visualize. Within education, it can be challenging to present abstract topics in simple, visual ways that resonate with students. Often, educators rely on images, textbooks, and software to help highlight core concepts.

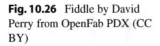

Fig. 10.26 Fiddle by David
Perry from OpenFab PDX (CC
BY)

Now with the introduction of 3D printing, educators, artists, scientists, engineers, and innovators all over the world can create physical models to complement and even enhance their understanding of mathematical methods [21–23].

10.6.1 Geometry

One entertaining and engaging way to teach basic geometry to younger students is through a castle building workshop! Students are tasked with designing and 3D printing a set of geometric shapes, which they can initially create using Autodesk Tinkercad (https://www.tinkercad.com/), a simple and lightweight browser-based introductory CAD program. We recommend that students initially prototype their castles with simple rectangles, squares, cylinders, cones, and polygons and then export each unique shape individually. This can be done by selecting one shape, navigating to *Export*, selecting "Include: the selected shape" and choosing an STL file. Next, students should determine how many of each unique shape to print to complete their castles. Finally, they can load each shape into Cura or their preferred slicer and 3D print the desired amount of shapes (Fig. 10.27).

Finally, students can use what they know about calculating surface area and volume and summing together the geometric castle parts to see how large of a footprint the castle may take up. There are plenty of online resources, such as https://www.thoughtco.com/surface-area-and-volume-2312247, that educators can use to provide students with helpful 'how-to' calculation guides. More advanced lessons for older students might consider 3D printing their own Lego bricks, designing them from scratch in a CAD program, or using calipers to measure precise dimensions.

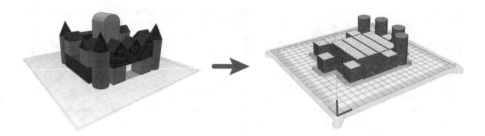

Fig. 10.27 Learning geometry may be easier to conceptualize when you 3D model and then calculate your design's surface area and volume. Screenshot from Ultimaker Cura (https://ultimaker.com) software

10.6.2 Calculus and Abstract Math

Several peer-reviewed papers detail how anyone can apply 3D printing to advanced calculus [24, 25]. In particular, 3D printing provides a rare opportunity for makers of all ages and abilities to develop their spatial math skills. When learning or teaching concepts such as solids of revolution, consider using a CAD program like Onshape (https://www.onshape.com/) to create 2D sketches and 3D prints. Sue Francis, a math educator at the American International School of Guangzhou, taught students how to use simple CAD programs to visualize how the 2D images would appear in 3D space [26]. Students were able to 3D model and then 3D print each of the shapes they made. Tasked with determining which had the highest volume, students could hold the shapes, sense weight, and manipulate the shapes in any way they wanted (Fig. 10.28).

In another example, students can grow more familiar with graphs of functions of two variables by 3D modeling the shape in powerful software such as Mathematica (https://www.wolfram.com/mathematica/). By simply entering the equation and establishing some parameters, students can create 3D models of complex objects. In Mathematica, we can follow the example below to create and export an example parametric surface as an STL:

```
f[u_, v_] := {u, v, u^2 - v^2};
scale = 40;
radius = 0.75;
numPoints = 24;
gridSteps = 10;
curvesU = Table[scale*f[u, i], {i, -1, 1, 2/gridSteps}];
curvesV = Table[scale*f[j, v], {j, -1, 1, 2/gridSteps}];
tubesU = ParametricPlot3D[curvesU, {u, -1, 1}, PlotStyle -> Tube[radius,
PlotPoints -> numPoints], PlotRange -> All];
```

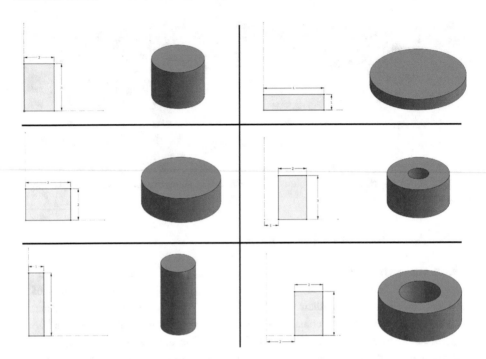

Fig. 10.28 Conceptualizing complex calculus can help students better visualize shapes that are revolved around the Y axis and determine which has the biggest volume. Lesson plan developed by Sue Francis at the American International School of Guangzhou

```
tubesV = ParametricPlot3D[curvesV, {v, -1, 1}, PlotStyle -> Tube[radius,
PlotPoints -> numPoints], PlotRange -> All];
corners = Graphics3D[Table[Sphere[scale f[i, j], radius], {i, -1, 1, 2}, {j, -1, 1,
2}], PlotPoints -> numPoints];
output = Show[tubesU, tubesV, corners]
Export["MathematicaParametricSurface.stl", output]
```

We can simply click the output and select "open file directory" to locate the export 3D model file. You can then import the 3D shape into Cura or your slicer of choice (Fig. 10.29) and 3D print it!

10.6.3 Mathematical Art

From fractals and golden ratios to patterns and symmetry, it's certainly no secret that math and art are closely linked. M. C. Escher (1898–1972) was a famous artist known for his

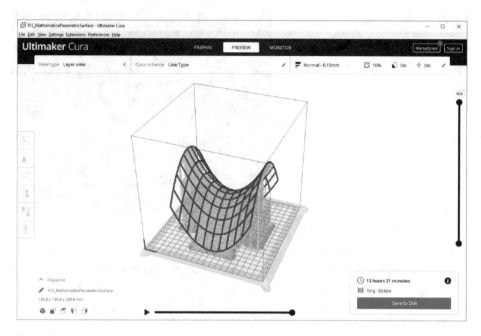

Fig. 10.29 To help better visualize concepts, it's possible to 3D model and then 3D print complex graphs of functions of two variables. Screenshot from Ultimaker Cura (https://ultimaker.com) software

geometric works of art. Though he denied an ability to perform or understand complex math, the proof is in his work. Initially inspired by Islamic patterns in the Alhambra—a fourteenth-century palace in Spain—Escher was fascinated by repeating patterns, tessellation, and the concept of infinity. He consulted mathematical textbooks and consulted with renowned mathematicians to create beautiful and world-acclaimed lithographs, woodcuts, wood engravings, and drawings (Fig. 10.30).

Fig. 10.30 M. C. Escher drew much of his creative inspiration from geometric art and pattern. Escher's "Swans (1956)" artwork. Photo by Pedro Ribeiro Simões (CC-BY-2.0)

Fig. 10.31 Stereographic
projection of Escher's "Lizard"
by Volker Schuller (CC BY-SA
3.0)

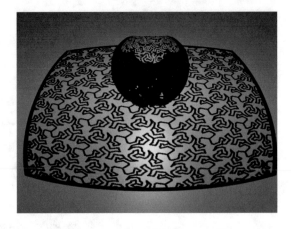

For those up for a genuine challenge, you might consider 3D printing a 3D stereographic projection of Escher's famous 1942 "Lizard" woodcut. You can think of stereographic projection as a way to show 3D patterns in 2D space. By shining a light through the top of the spherical model, the curves of the Escher 3D print create shadows and display the 3D sphere as a linear 2D plane. In addition to innovative art applications, stereoscopic projection is frequently used in fields such as geology, cartography, and photography.

The 3D model (Fig. 10.31) was created by maker and designer Volker Schuller (CC BY-SA 3.0) using OpenSCAD (https://openscad.org/). You can download the stereographic projection 3D model by visiting Schuller's page at https://www.thingiverse.com/thing:428826. When 3D printing, ensure you're carefully considering infill and—most importantly—support settings. Because of the complexity of this 3D object, removing supports can be a slow process but worth it in the end!

Finally, what if you want to break the laws of physics? With 3D printing, you can explore the art of beautiful fractals, abstract geometry, and even impossible shapes (Fig. 10.32) by creating your own objects, such as:

- A fantastic fractal Sierpiński Pyramid: https://www.thingiverse.com/thing:1356547 designed by Yi-Ting Tu (CC BY 4.0)
- An impossible Penrose triangle: https://www.thingiverse.com/thing:547580 designed by Kanbara Tomonori (CC BY 4.0)
- A four-dimensional Klein Bottle: https://www.thingiverse.com/thing:145694 designed by Robert Woodhead (CC BY 4.0)
- A singled-sided Mobius strip: https://www.thingiverse.com/thing:239158, designed by Chris Wallace (CC BY-SA 3.0)
- An impossible cube: https://www.thingiverse.com/thing:4764837 designed by Martin Anderson-Clutz (CC BY 4.0)

Fig. 10.32 Trending left to right, starting at the top: A 3D printed Sierpiński Pyramid, Penrose triangle, Klein bottle, and impossible cube. Designers credited on the right

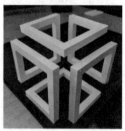

10.7   Summary

The goal of this chapter was to provide you with a diverse range of 'STEAM-building' exercises that might ignite an innovative spark in a variety of different interests and projects. We started first in the sciences, and explored how you can apply 3D printing to extract meaningful information on diet, behavior, and size simply by looking at jaws and teeth. Next, we discussed how 3D printers are changing the game when it comes to custom medical implants. We learned how to 3D print custom prosthetics and how to cater different designs to different users—including possibly even adding a second thumb to each hand! Next, we took a deep dive into accessible, affordable robotics through the open-source Otto robot, where we reviewed how to use 3D printing to augment the initial design and how to begin coding the robot to walk, jump, and dance. Next, we explored the world of 3D printed musical instruments, with some beautiful-sounding and perhaps not-so-beautiful sounding examples. Last, we reviewed several ways that math and 3D printing go hand-in-hand, with educational applications as well as abstract, artistic mathematical expression.

10.8   Chapter Problems

- Can you think of additional science experiments that might be well complemented by 3D printing?
- How might you adjust the prosthetic files to add an attachment to hold something unique?

- How might you adjust the prosthetic files for a user with a bigger forearm? What considerations would you have to make?
- What types of additional devices would you want to integrate into a prosthetic?
- How can you modify the Otto robot to dance left to right? What if you want to make it go faster or slower?
- Have you tried 3D printing several of the instruments? Which sounds the best? The worst? Why might that be?
- What are some of the K-12 applications to teach math through 3D printing? What types of additional lesson plans can you think of that can include 3D printed parts?

References

1. Kerr, T., Patrick, K., Clementz, M. T. & Vietti, L. No Fossils Were Harmed During the Training of This Preparator: Using 3D Models to Teacher Proper Preparation Techniques. in The Society of Vertebrate Paleontology 77th Annual Meeting (2017).
2. Cavigelli, J.-P. et al. A Test of the State of the Art in 3D Color Printing. in The Society of Vertebrate Paleontology 80th Annual Meeting (2020).
3. Johnson, E. H. & Carter, A. M. Defossilization: A Review of 3D Printing in Experimental Paleontology. Frontiers in Ecology and Evolution vol. 7 Preprint at https://doi.org/10.3389/fevo.2019.00430 (2019).
4. Peterson, J. E. & Krippner, M. L. Comparison of Fidelity in the Digitization and 3D Printing of Vertebrate Fossils. Journal of Paleontological Techniques 22, 1–9 (2019).
5. Lewis, D. The fight for control over virtual fossils. Nature (2019).
6. Naturalis Biodiversity Center. Naturalis Biodiversity Center homepage. https://www.naturalis.nl/en.
7. Mussetter, B., Hendrickson, M. & Perez, V. How Big Was the Megalodon? Lesson Plan. iDigFossils https://www.idigfossils.org/wp-content/uploads/2019/03/Mussetter_Megalodon-Lesson-Plan_4th-Grade.pdf.
8. Manero, A. et al. Implementation of 3D Printing Technology in the Field of Prosthetics: Past, Present, and Future. Int J Environ Res Public Health 16, (2019).
9. Liacouras, P. C. et al. Using computed tomography and 3D printing to construct custom prosthetics attachments and devices. 3D Print Med 3, (2017).
10. Ventola, C. L. Medical Applications for 3D Printing: Current and Projected Uses. P T. 39, 704–711 (2014).
11. e-NABLE. Enabling the Future homepage. https://enablingthefuture.org/.
12. Kieliba, P., Clode, D., Maimon-Mor, R. O. & Makin, T. R. Robotic hand augmentation drives changes in neural body representation. Sci Robot 6, 1–29 (2021).
13. Clode, D. The Third Thumb web page. https://www.daniclodedesign.com/thethirdthumb.
14. Arduino homepage. https://www.arduino.cc/.
15. Arduino. Arduino Project Hub web page. https://create.arduino.cc/projecthub/.
16. OttoDIY. OttoDIY homepage. https://www.ottodiy.com/.
17. OttoDIY. OttoDIY file directory web page. Wikifactory https://wikifactory.com/+OttoDIY/otto-diy.

18. Kantaros, A. & Diegel, O. 3D printing technology in musical instrument research: reviewing the potential. Rapid Prototyping Journal vol. 24 1511–1523 Preprint at https://doi.org/10.1108/RPJ-05-2017-0095 (2018).
19. Avanzini, F., Barate, A. & Ludovico, L. A. Digital Fabrication: 3D Printing in Pre-School Education. Open and Interdisciplinary Journal of Technology 14, 71–92 (2018).
20. Dabin, M., Narushima, T., Beirne, S. T., Ritz, C. H. & Grady, K. 3D Modelling and Printing of Microtonal Flutes. https://ro.uow.edu.au/lhapapers/2798 (2016).
21. Segerman, H. Visualizing Mathematics with 3D Printing. https://www.3dprintmath.com/.
22. Huleihil, M. 3D printing technology as innovative tool for math and geometry teaching applications. in IOP Conference Series: Materials Science and Engineering vol. 164 (Institute of Physics Publishing, 2017).
23. Sun, Y. & Li, Q. The Application of 3D Printing in Mathematics Education. in The 12th International Conference on Computer Science & Education 1–4 (2017).
24. Paul, S. 3D Printed Manipulatives in a Multivariable Calculus Classroom. PRIMUS 28, 821–834 (2018).
25. Dilling, F. & Witzke, I. The Use of 3D-Printing Technology in Calculus Education: Concept Formation Processes of the Concept of Derivative with Printed Graphs of Functions. Digital Experiences in Mathematics Education 6, 320–339 (2020).
26. Shaw, M. & SteamHead. 3D Printing for Calculus - Sue Francis Workshop on Math Visualization. https://www.youtube.com/watch?v=O63kPj3TEeI (2018).

Resources

3D printers:

- Prusa 3D printers: https://www.prusa3d.com/
- Ultimakers: https://www.ultimaker.com/
- Creality: https://www.creality.com/
- LulzBot: https://www.lulzbot.com/
- Flashforge: https://www.flashforge.com/
- Monoprice: https://www.monoprice.com/pages/3d_printers

Slicer software:

- Ultimaker Cura slicer software: https://ultimaker.com/software/ultimaker-cura
- PrusaSlicer: https://www.prusa3d.com/page/prusaslicer_424/
- OctoPrint: https://www.octoprint.org/
- Slic3r: https://www.slic3r.org/

Thermoplastic material guides:

- Prusa Material Table: https://www.help.prusa3d.com/materials
- Ultimaker material guide: https://www.support.ultimaker.com/hc/en-us/categories/360 002336619-Materials
- 3D Printing Materials—The Ultimate Guide: https://www.all3dp.com/1/3d-printing-materials-guide-3d-printer-material/
- MatterHackers Material Guide: https://www.matterhackers.com/3d-printer-filament-compare

© The Editor(s) (if applicable) and The Author(s), under exclusive license
to Springer Nature Switzerland AG 2022
T. Kerr, *3D Printing*, Synthesis Lectures on Digital Circuits & Systems,
https://doi.org/10.1007/978-3-031-19350-7

3D printing troubleshooting guides:

- All3DP: https://www.all3dp.com/1/common-3d-printing-problems-troubleshooting-3d-printer-issues/
- MatterHackers: https://www.matterhackers.com/articles/3d-printer-troubleshooting-guide
- Simplify3D: https://www.simplify3d.com/support/print-quality-troubleshooting/
- Prusa: https://www.help.prusa3d.com/category/troubleshooting_194
- Ultimaker: https://www.support.ultimaker.com/hc/en-us

3D printing applications:

- Arduino Otto instruction guides, https://www.create.arduino.cc/projecthub/
- Otto DIY robot 3D files and instructions, https://www.ottodiy.com/
- Advanced Otto instructions, https://www.wikifactory.com/+OttoDIY/otto-diy

Index

A
Adaptive layers, 97
Adding or fusing, 2
Additive manufacturing, 2
Aerospace Industry, 19
Anisotropic, 29
Anisotropy, 29
Applications, 11
    aerospace industry, 19
        international space station, 19
        lunar habitats, 19
        Martian habitats, 19
        zero-gravity 3D printing, 19
    artistic reconstruction, 15
    automotive industry, 18
    bioprinting, 15, 16
    bone implants, 17
    bridges, 18
    clothing, 14
    construction, 17
    constructionist learning, 14
    dentistry, 16
    education, 13
    fashion industry, 14
    food, 15
    healthcare, 139
    housing, 17
    jewelry, 14
    manufacturing, 12
    meat, 15
    nutrition, 16
    organs, 16
    prosthetics, 16
    rapid prototyping, 12
        rapid tooling, 12
    research, 14
    stem cells, 16
    surgery, 16
    sustainable prototyping, 14
    tissue, 16
Arduino, 144
Art, 14
Automotive Industry, 18
Avoid supports when traveling, 90

B
Bed adhesion, 40
    brims, 40
    rafts, 41
    skirts, 40
Bowden, 39
Bowden extruders, *see* 3D printer anatomy
Bowden tube, *see* Bowden extruders
Brim line count, 95
Brim width, 95
Brims, 40, 73, 94
Build plate, *see* 3D printer anatomy
Build plate adhesion, 75
Build plate temperature, 87
Build plate temperature initial layer, 87

C
CAD software, 65
Cartesian, 36

Cold end, *see* 3D printer anatomy
Combing mode, avoid printed parts when
    traveling, 90
Computer-aided design, *see* CAD software

D
Delta, 36
Desktop manufacturing, 3
Direct drive extruders, *see* 3D printer anatomy

E
Education, 13
Enable conical support, 97
Enable print cooling, 91
Enable retraction, 90
Enable support interface, 93
Extrude, 2
Extruders, *see* 3D printer anatomy

F
Fan speed, 91
Fashion, 14
FDM 3D printers, 36
    Cartesian, 36
    Delta, 36
Filament, 2
    filament spool, 43
    load filament, 124
    unload filament, 126
Filament spool, 43
Final printing temperature, 87
Food, 15
Fused deposition modeling, 5
Fused filament fabrication. *See also* Fused
    deposition modeling

G
G-code, *see* slicer
Generate supports, 92
Gradual infill steps, 87
Gradual support infill steps, 93

H
Healthcare, 16

Heat block, 42
Heat creep, 42, 124
Heat sink, 42
Horizontal expansion, 84
Hot end, *see* 3D printer anatomy
Housing, 17

I
Infill, 75
Infill density, 86
Infill patterns, 86
Initial fan speed, 91
Initial layer print speed, 89
Initial printing temperature, 87
Ironing, 85
Isotropy, 29

J
Jewelry, 14

L
Layer height, 74, 83
Lift head, 91
Line width, 84

M
Make overhang printable, 96
Manufacturing, 12
Material, 70
Material jetting, 8
Materials
    material flexibility, 27. *See also*
        thermoplastic
Maximum model angle, 96
Minimum layer time, 91
Minimum speed, 91
Mold, 96
Move, 68

N
Nozzle, 43
Nutrition, 15

P

Photopolymer. *See also* resin
Polylactic acid (PLA), 31
Primer blob, 126
Primer line, 126
Printer, 70
Printing temperature, 87
Printing temperature initial layer, 87, 88
Print sequence, 95
Print speed, 89
Profile dropdown, 81
Project file, 76
Prosthetics, 16, 139
Prusa 3D printer anatomy
   direct drive extruder, 54
   extruder fans, 57
   heat block, 56
   heated, flexible, removable magnetic build
    plates, 54
   heat sink, 56
   hobbed bolt, 57
   hot end assembly, 54
   nozzle, 56
   PINDA probe, 56
   x-axis, 54
   y-axis, 54
   z-axis, 54. *See also* 3D printer anatomy

Q

Quality settings, 70

R

Rafts, 95
Rapid prototyping, 12
Regular fan speed at height, 91
Regular fan speed at layer, 91
Research, 13
Resin, 6, 8
Retraction distance, 90
Retraction speed, 90
Rotate, 68
Robotics, 144

S

Scale, 69
Selective laser sintering, 7

Setting categories, 82, 83
Setting parameters, 82
Setting profile, 82
Setting search bar, 81
Setting visibility menu, 82, 83
Sintering, 7
Skirt, 40, 94
Skirt/brim speed, 89
Skirt line count, 94
Slicer, 37. *See also* slicing software
Slicer settings, *see* Ultimaker Cura
   bed adhesion
    brim, 94
    raft, 95
    skirt, 94
   brim, *see* bed adhesion
   build plate temperature, 87
    build plate temperature initial layer, 87
   cooling, 91
    fan speed, 91
    initial fan speed, 91
    regular fan speed at height, 91
   experimental settings, 96
    adaptive layers, 97
   importing files, 68
   infill, 71, 86
    gradual infill, 87
   infill density, 86
   infill pattern, 86
    3D concentric infills, 86
    3D infills, 86
    quick 2D infills, 86
    strong 2D infills, 86
   initial layer height, 83
   ironing, 85
   layer height, 70
   line width, 84
   navigation, 68
   printing temperature, 87
    printing temperature initial layer, 87
   print speed, 89
    initial layer print speed, 89
    skirt/brim speed, 89
   quality settings, 70
   resolution, 70
   retraction, 90
    retraction distance, 90
    retraction speed, 90
   special mode settings, 95

supports, *see* supports. *See also* supports
   support density, 93
   support overhang angle, 92
   support pattern, 92
   travel speed, 89
   wall line count, 84
   wall thickness, 84. *See also* Bed adhesion
Slicing software, 37
Spiralize outer contour, 96
STEAM, 129
STEAM Projects, 129
   anthropology, 129
   archaeology, 129
   art, 149
   comparative anatomy, 132
   engineering, 144
      "Otto" Arduino robot, 144
      arduino, 144
      arduino IDE
      robotics. *See also* robotics
   Math, 155
      Calculus and Abstract Math, 157
      Mathematical Art, 158
      Visualizing geometry, 156
   museums, 130
   musical instruments, 149
      percussion, 153
      string instruments, 154
      woodwinds, 149
   paleontology, 129
      archaeocetes, 132
      British Museum, 130
      Cats and Dogs, 135
      D'Arcy Thompson Zoology Museum, 133
      3D scanning, 130
      diet and behavior, 133
      fossils, 130
      fossil whales, 131
      Idaho Virtualization Laboratory, 130
      London Natural History Museum, 130
      Mace Brown Museum of Natural History, 133
      Megalodon, 137
      Mysticetes, 134
      Naturalis Biodiversity Center, 130
      Smithsonian, 130
      teeth, 137
      University of Dundee, 130

University of Edinburgh, 133
University of Wyoming, 130, 136
prosthetics. *See also* Prosthetics
   Dani Clode. *See also* Prosthetics
   e-NABLE, 140
   Third Thumb project. *See also* Prosthetics
Science, 129
Technology, 139. *See also* Arduino
Stereolithography, 6
   galvanometers, 6
Supports, 40, 75
Support density, 93
Support horizontal expansion, 93
Support infill layer thickness, 93
Support overhang angle, 92
Support pattern, 92
Support placement, 92
Support structure, 92
Surface mode, 96

T
Thermoplastic, *see* filament
   acetonitrile butadiene styrene (ABS), 60
   materials guide, 88
   polyethylene terephthalate glycol (PETG), 60
   polylactic acid (PLA), 59
   specialty filaments, 61
   thermoplastic polyurethane (TPU), 61
Troubleshooting, 98
   clog, 110
   filament grinding, 113
   jam, 110
   layer shift, 104
   nozzle scraping, 117
   peeling, 100
   ringing, 109
   stringing, 89
   stringing, oozing, 102
   support scarring, 107
   under-extrusion, 106
   warping, 100
3D printer anatomy, 41
   belts, 41
   heat block, *see* extruders
   heat break, *see* extruders
   heat sink, *see* extruders

linear rails, 41
linear rods, 41
nozzle, *see* extruders
thermistor, *see* extruders
x-axis, 41
y-axis, 42
z-axis, 41
3D printer brands, 46
    Creality, 50
        Creality Ender-3 S1 Pro, 50
        Creality Ender-3 V2, 50
    Flashforge, 51
        Flashforge Creator Pro 2, 52
        Flashforge Finder, 51
    LulzBot, 50
        LulzBot SideKick, 51
        LulzBot TAZ 6, 51
    Monoprice, 52
        Monoprice MP Cadet, 52
    Prusa Research, 46
        Prusa i3 MK3S+, 46
        Prusa MINI+, 46
    Ultimaker, 48
        Ultimaker S3, 48
        Ultimaker S5, 49

        Ultimaker S5 Pro, 49
Travel speed, 89
3D scanning, 14, 130

U
Ultimaker Cura, *see* slicer

V
Vase mode, 96

W
Wall line count, 84
Wall thickness, 84

Y
Y-axis, 42
Y-H-T rule, 73

Z
Z-hop when retracted, 90

Printed in the United States
by Baker & Taylor Publisher Services